T0363097

Gemstones
IN VICTORIA

William D. Birch & Dermot A. Henry

MUSEUMSVICTORIA

PUBLISHING

This edition published in 2013
Reprinted 2017, 2021, 2022 by
Museums Victoria Publishing
11 Nicholson Street
Carlton, Victoria 3053, Australia
publications@museum.vic.gov.au
www.museumsvictoria.com.au

First published in 1997 by The Royal Society of Victoria and
The Mineralogical Society of Victoria as Special Publication No. 4 of
The Mineralogical Society of Victoria.

 A catalogue record for this
book is available from the
National Library of Australia

ISBN 9781921833069

Design by studioether
Printed in China by RR Donnelley Asia Printing Solutions Ltd.

5 7 9 10 8 6 4

Museums Victoria acknowledges the Wurundjeri and Boon Wurrung peoples of the Kulin Nations where we work, and First Peoples language groups and communities across Victoria and Australia. Our organisation, in partnership with the First Peoples of Victoria, is working to place First Peoples living cultures and histories at the core of our practice.

The Authors

◇◇◇◇◇◇◇◇◇◇◇◇◇◇◇◇◇◇◇◇◇◇

Dr William Birch

Senior Curator in Geosciences at Museums Victoria

Dr Birch was appointed Curator of Minerals in the then National Museum of Victoria in 1974. He has been responsible for the growth and enhancement of the Museum's collections of minerals, rocks, ores, gemstones and meteorites. His research and collecting efforts have taken him to many far-flung places in Australia and overseas, but he has retained a focus on Victoria because its complex geology provides a great variety of interesting rocks and minerals. Dr Birch has described nearly 40 new minerals and published over 200 research papers and articles, in addition to editing six books. His involvement with geological and mineralogical organizations in Australia and overseas has been recognised through the award of the Selwyn Medal by the Geological Society of Australia in 1999 and an AM (Member of the Order of Australia) in 2006.

Dermot Henry

Manager of Natural Science Collections at Museums Victoria

Dermot Henry commenced at Museum Victoria in 1982 and worked in a number of management roles within the geological collections, before his appointment as Manager of Natural Sciences collections in 2001. He has published articles on a variety of topics covering minerals, rocks and meteorites and has contributed to three books on Victorian minerals. Since 1995 he has been the editor of the *Australian Journal of Mineralogy*. He is a keen field collector of rocks and minerals for the Museum, collecting at sites around Australia and overseas. He has also obtained significant donations for the Museum through the Federal Government's Cultural Gifts program. Dermot was also instrumental in developing the themes and content for the *Dynamic Earth* exhibition, which opened in 2010 at Melbourne Museum.

Preface

◇◇◇◇◇◇◇◇◇◇◇◇◇◇◇

When all copies of the first edition of this book sold out in 2009, Museums Victoria offered to publish a revised version, with the agreement of both the Mineralogical Society and the Royal Society, the original joint publishers. The book has been redesigned and updated to include information on new discoveries and many additional images of faceted gemstones to augment the text. We have kept the arrangement of the descriptions of Victorian gemstones substantially the same as the original version, so as to provide historical, regional and geological contexts for each class of gemstone. A section at the end of each chapter has been added, providing brief descriptions of selected gemstone occurrences in other Australian states.

As is immediately obvious, Victoria is far from being a region devoid of gemstones. It's just that, in general, they are not as prolific or valuable as those found in other Australian gemfields, and many localities are now exhausted or out-of-bounds to collectors. This edition, like the original, is not intended as a guide to gem fossicking, but attempts to marry historical and geological information in ways that not only explain why gemstones occur where they do, but also might assist in the discovery of new occurrences. A spin-off from these two editions has been the renewed emphasis on acquiring gemstones for Museums Victoria's collections, from both old collections and recent discoveries. Long-term preservation of these specimens will ensure they are available for future research into the geological processes that have shaped Victoria.

As the content of the first edition has largely been retained, those individuals who contributed specimens and information remain significant and relevant. Rather than list them all again here, the original Acknowledgments page has been reproduced at the back of this edition. We note with some sadness that quite a few of those generous people have since passed away.

For this edition, we are grateful to the many individuals, institutions and companies for providing new information and specimens for photography. Without their support this book would not have been possible. We'd also like to thank Patty Brown, Publishing Manager Museums Victoria Publishing.

Bill Birch and Dermot Henry

Contents

◇◇◇◇◇◇◇◇◇◇◇◇◇◇◇◇◇◇

Discovery

"…as yet no one country on the broad earth has yielded such an assemblage of varieties of rare and precious gems as Victoria."

REV. JOHN BLEASDALE, 1868

Victoria is often the forgotten State when it comes to gem minerals. Most books and articles on the precious and semi-precious stones of Australia at best recycle outdated and vague information on Victorian localities. Gemstone maps often show Victoria as a blank space. This is quite a contrast to earlier times when exciting discoveries were being made, and the Reverend Bleasdale felt entitled to claim a unique place in the world for Victorian gem minerals.

Discovery

The early history of gem minerals in Victoria is inextricably linked to the official discovery of gold in 1851 and the rushes that followed. It was inevitable that a few out of the host of diggers sifting the rich alluvial deposits for gold would be curious enough to save the small, sparkling, brightly coloured crystals often found in the residues from their gold-pans and sluice-boxes. Probably because small quartz crystals were easily mistaken for diamonds they, along with assorted coloured stones, were sent to various jewellers in Melbourne for examination. In a colony eager for mineral wealth of any sort, the identification of some stones as sapphires and zircons soon generated some rather grandiose predictions about a future gem industry.

The first colonial identity to publicise the discovery of gem minerals in Victoria was George Milner Stephen, barrister and public servant. Stephen addressed a meeting of the Geological Society of London in 1854, and illustrated his talk with gems and gold crystals from the Australian colonies. His display included a blue and white waterworn sapphire from the Ballarat diggings, as well as zircons, 'spinel ruby', tourmaline, pale-yellow topaz and garnet from the River Ovens, in the northeast of Victoria. Stephen surmised that the gems occurred in what were ancient river courses, thereby explaining their waterworn appearance. In 1857, the German natural historian Ludwig Becker reported small gemstones such as topaz, sapphires, rubies and garnets occurring with gold and other heavy minerals from various gold diggings in the colony.

Unfortunately, few of the earliest discoveries were recorded in detail. Amongst the first was a report in the [Melbourne] *Herald* of 28 July,

1860, which reprinted an article from the *Mount Alexander Mail*. The subject of the note was a description by the government geologist, George Ulrich, of a matchbox full of small precious stones. These had been collected by a prospector washing drift near Guildford, on the Loddon River. Ulrich found most of the stones were sapphires, mainly blue, but with several greenish stones, one rose-coloured 'ruby', and some transparent, coloured zircons. He drew an analogy with similar gems found in volcanic areas in Germany and France and predicted that the Guildford stones came from 'old basalt higher up the Loddon'. Ulrich urged miners to closely examine the washings in their tin dishes in the hope of discovering more 'brilliant gems'. Even the white shiny grains should be preserved, because *"according to reports, a diamond has been found at Yandoit, where . . . the geological character of the rocks is not different from that at the Loddon"*.

The Ovens District, as the river system and its goldfields had come to be known, quickly gained fame as Victoria's richest gemfield. This was no doubt sparked by the discovery of small diamonds in the gold-bearing wash from the district's Woolshed Valley, north of the town of Beechworth, in the early 1860s. Even so, there were few systematic attempts made to exploit the gems, and only a few Melbourne jewellers, such as George Crisp (of Queen Street), James Murray (of Bourke Street) and James Spink and Son (of Collins Street), actually cut some stones. This apparent lack of interest concerned the Reverend Bleasdale, who set out on a personal crusade to collect and promote the precious stones of the Colony. His paper read to the Royal Society of Victoria in November 1863 described diamonds, sapphires, zircons, topaz, and quartz varieties from Victoria. It was illustrated with

◊ **1.1** Deep lead dumps near Guildford (1996).

a display featuring gems from his own collection and some borrowed from friends. Bleasdale placed the best stones he could obtain from Ceylon, the East and West Indies, Brazil and Peru side-by-side with the Victorian specimens, so their quality could be compared. The 'gem of the evening' was Mr Murray's cut Beechworth diamond, weighing a little less than 2 carats.

Bleasdale maintained his enthusiasm for Victorian gem minerals. In October 1864 he proposed that the Royal Society sponsor an exhibition of colonial gems and jewellery in Melbourne. Such an exhibition, he argued, would promote the Society's good works and reputation amongst a 'somewhat sceptic public'. It would help to elevate public taste in the fine arts and encourage the use of colonial materials. Also, with a moderate admission fee, a multitude of visitors would derive a handsome profit for the Society. The exhibition was held during the week ending 6 May, 1865, in the Royal Society's Hall, just as Bleasdale assumed the presidency of the Society. He had succeeded in attracting exhibits of Victorian and foreign gems and gold and silver jewellery from the most prominent Melbourne jewellers and private collectors, to add to his own display. Beechworth gemstones — notably six diamonds, one of which was mounted uncut in a ring — featured prominently in a display by William Turner. George Stephen contributed over 250 gems, including

diamonds, sapphires of many kinds, topazes, garnets and zircons, mostly from the Ovens District and Ballarat. Stephen's most spectacular stone was a 17.9-carat green sapphire, 1.75 inches (4.4 cm) across the prism, from Jim Crow Creek, on the Daylesford goldfield. Many of Stephen's Beechworth gemstones had been given to him by Edward Dunn, whom Stephen had sought out when he visited Beechworth early in 1861. Dunn, only 17 at the time, had acquired a fine collection of local gemstones but was unable to identify many of them. In return for supplying Dunn with names, Stephen was given any stones he wanted.

A detailed report on the Royal Society's exhibition, describing the main colonial gem minerals and providing a catalogue of exhibitors, was presented to the Society by Bleasdale in late 1865. Bleasdale's authority on the gem minerals was acknowledged by the publication of his essay 'Gems and precious stones found in Victoria' in the catalogue of the Intercolonial Exhibition in Melbourne in 1866–7. This was a companion-piece to a more extensive essay on Victorian mineral species by George Ulrich. Bleasdale's contribution was based heavily on his earlier report on the Royal Society exhibition and, together with Ulrich's notes, summarised and updated information on the well-known species, such as diamond, corundum, spinel, zircon and topaz, as well as some of the less familiar

gem materials, such as pearls and semi-precious stones, on display in the exhibition. No doubt in support of his cause of promoting local gem stones, Bleasdale quoted correspondence from Professor Maskeleyne, at the British Museum, extolling the virtues of Victorian gems. He also took a swipe at popular taste in jewellery, lambasting the 'tawdry trifles' and 'trashy articles' seen everywhere on a 'bedizened' populace. He reserved special disdain for pieces in the form of a *"tawdry impossible abortion of a leaf unknown to botanists, badly engraved, and rendered ridiculous by a wretched imitation of a ruby or emerald, placed where nature never placed either ornament or fruit."*

In July 1867 Bleasdale again used the forum of the Royal Society to report the discovery of a new gem mineral from Victoria. He displayed six specimens of what he referred to as transparent red tourmaline, rubellite, found as small crystals enclosed in quartz crystals from the 'Broadford Lead', Tarrangower (the locality is the Bradford Lead, near Maldon). Bleasdale stated that the Chemist at the Department of Mines, James Cosmo Newbery, had examined the crystals and was satisfied with their identification. These were subsequently identified as red garnets by George Ulrich, who described them in some detail in his landmark 1870 publication Contributions to the Mineralogy of Victoria. In the same article, Ulrich described new sapphire and ruby discoveries in William Wallace Creek, near Berwick. He had firsthand knowledge of this locality, because he had accompanied Bleasdale and Alfred Selwyn, the Director of the Geological Survey, on a wagon trip to the creek, via the newly opened Bowmans Track, in 1868. According to Bleasdale's account of the journey, published in the *Colonial Monthly* in October 1868, the three men had 'walked up the creek and panned a few dishes of wash-dirt, turning up cassiterite, agates, chalcedony and quartz crystals'. Bleasdale was instrumental in arranging displays of colonial gems in the great International Exhibitions of the day, such as those in Melbourne in 1872–3 and 1875, and in Philadelphia in 1876. He wrote notes updating new finds of diamonds, sapphires and other gemstones to accompany the displays, which featured many stones from his own collection.

The potential of Victoria as a gemfield was given official recognition by the Mining Department. Between 1865 and the early 1880s, the Annual Reports of the Secretary for Mines included a summary of discoveries of diamonds, sapphires and other gem minerals, based mainly on reports from the Mining Wardens in the goldfields. Diamonds were considered the most significant gem, so much so that a running total of discovered stones was kept from 1864 until 1884.

As most of the easily won alluvial gold deposits were exhausted and abandoned, reported discoveries of gem minerals petered out. Enthusiasm dwindled as Bleasdale, Ulrich and Dunn, who had all championed the cause of Victorian gemstones, left the colony. There was a flurry of excitement in 1912 when diamonds were found in the Great Southern deep lead gold mine near Rutherglen, but this was short-lived. Any flickering interest in Victoria as a gem mineral producer was snuffed out

as the rich opal and sapphire fields in other states such as South Australia, New South Wales and Queensland were discovered.

After the first discoveries, one hundred years elapsed before interest in Victorian gemstones was rekindled. This time it was based on the new hobby of gemstone fossicking and amateur lapidary. The first gem club formed in 1959 under the wing of the Victorian Branch of the Gemmological Society of Australia. The club's enthusiasm for collecting was a natural spin-off from the Society's scientific studies of gem materials. The Lapidary Club of Victoria was founded in 1962, to be followed in the next decade or so by scores of similar clubs in Melbourne and country Victoria. During the 1960s and 1970s, many of the alluvial gravels and old deep lead dumps in the long-abandoned goldfields provided a source of material for polishing and faceting. The intense collecting activity uncovered many new locations for gem minerals. As in the past, however, the interest in local gemstones lost momentum as the more productive localities became exhausted or less accessible towards the end of the 1970s. Today, while many of the clubs survive, their activities are more diverse and less dependent on collecting local material on field trips.

The enigma of the Beechworth and Rutherglen diamonds has lured various mining companies, even to the present day, but despite the use of modern prospecting methods, no new discoveries have been reported.

For all the grand predictions of the 1860s, no viable gemfields have ever been established in Victoria. But this does not detract from the geological significance and scientific interest of the gem minerals, or from the pleasure to be had from finding them.

1.2 The Very Reverend John Ignatius Bleasdale (ca. 1870). La Trobe Picture Collection, State Library of Victoria.

1.3 Waterworn sapphire crystal (2.1 cm) obtained from Bleasdale by the Natural History Museum, London, in 1873.

The main players

REVEREND JOHN BLEASDALE

John Ignatius Bleasdale was born in Lancashire, England, in 1822. After a religious education and five years as a military chaplain, he transferred to the newly formed diocese of Victoria in 1850. Here he was able to combine his early ecclesiastical appointments in the Catholic Church with his deep interest in science. He became a prominent and well-respected member of many of the early scientific societies, especially the Royal Society of Victoria, and was a strong advocate for the establishment of the National Museum in 1854. Bleasdale wrote and spoke on a wide range of subjects, with his two great loves being colonial gems and wine. Where or from whom he gained his interest and knowledge in gemmology is not known, but throughout the 1860s and the early 1870s he drew public attention to the richness and diversity of Victoria's gem minerals. Apart from his journey to William Wallace Creek in 1868, in the company of George Ulrich and Alfred Selwyn, he would rarely have travelled into the countryside to visit the sites of gemstone discoveries. Nevertheless, through his enthusiasm and wide contacts he was able to amass what must have been a fine personal collection. He used it to great effect in the exhibition he organised for the Royal Society in 1865, but it received its ultimate recognition with an award at the Philadelphia Exhibition in 1876. During this period, Bleasdale served on many boards and committees dealing with public and religious education. His professional career culminated in 1874 when he was appointed Chancellor of the Archdiocese of Melbourne. Over the next few years, his health failed rapidly and in 1877 he resigned and moved to California, where he contributed his expertise to the developing wine industry. However, without a regular income he fell on hard times and died in 1884 in San Francisco. The fate of his substantial collection of Victorian gemstones is unknown; perhaps all that survives from his endeavours is a small collection he exchanged with the Natural History Museum in London in 1873.

GEORGE MILNER STEPHEN

It is difficult to summarise the quite remarkable and controversial career of George Milner Stephen, as it embraced the law, public service, land speculation, painting, politics, gold mining, gem collecting and, finally, faith-healing and spiritualism. Stephen was born in Somerset, England, in 1812 and emigrated to Sydney with his father in 1824. After completing his schooling in brilliant fashion, he was appointed to a legal clerkship in Hobart in 1830. However, with his career prospects uncertain, Stephen took advantage of some confusion over an offer to his brother and was appointed Advocate-General and Crown Solicitor in Adelaide in 1838. Despite some worthwhile achievements in public office, controversy dogged him over his land speculation, although subsequent legal proceedings saw him acquitted of perjury charges. After marrying in 1840, he visited Europe, where he courted noblemen by painting their portraits. Controversy again embroiled him after his return to Adelaide in 1846, with more court cases and suspect claims of gold and iron discoveries. After gold was discovered in Victoria, Stephen moved to Melbourne, where he began to combine his legal career with visits to the goldfields. During another visit to Europe between 1853 and 1855, he addressed the Geological Society of London on the subject of Victorian gemstones and donated Australian gems to the Dresden Museum, the latter gesture apparently made in the hope of securing some imperial honour. On his return to Melbourne in 1856, Stephen resumed his goldfield travels

and continued to build his collection of gem minerals. Further legal proceedings erupted over his previous perjury charges, but these could not prevent him being elected to Parliament as the member for Collingwood between 1859 and 1861. Stephen used the forum of Parliament to announce the discovery of diamonds in Victoria, laying a stone 'on the table' in May 1860. While pursuing a legal matter in Beechworth in 1861 and again in 1864, he met the teenage Edward Dunn, and benefited considerably from Dunn's generosity with his local collection of gemstones. Stephen was an enthusiastic participant in the Royal Society exhibition of colonial gemstones

◊

1.4 George Milner Stephen (1876). La Trobe Picture Collection, State Library of Victoria.

◊

1.5 Edward John Dunn (1888).
La Trobe Picture Collection,
State Library of Victoria.

1.6 *(opposite)*
George H. F. Ulrich (c. 1870).
La Trobe Picture Collection,
State Library of Victoria.

organised by the Reverend John Bleasdale in 1865 and in the Intercolonial Exhibition of 1866–7, for which he was awarded a medal. After a short stay in Beechworth, Stephen moved to Sydney, where he addressed the Royal Society of New South Wales in 1872 on the subject of Australian gems. In the late 1870s, he became convinced of his powers of faith-healing and gave up his law career to treat his patients for a variety of afflictions. As usual, controversy followed his claims. His powers were not enough to cure his own illness and he died in Brunswick, Melbourne, in 1894. Stephen was one of the strongest advocates for colonial gemstones, but the fate of his collection remains unknown.

EDWARD JOHN DUNN

Edward John Dunn's pioneering geological achievements in Australia and South Africa have not received the recognition they deserve. He was born in Bedminster, near Bristol, England, in 1844. His family emigrated to New South Wales in 1849 before settling at Beechworth, Victoria, in 1856. From his childhood, Dunn was fascinated by minerals and rocks. He roamed widely over the thriving alluvial gold diggings of the Woolshed Valley, collecting beautiful pebbles and visiting miners who saved gemstones for him from their wash-dirt. In 1860, he commenced a surveying course, but maintained his interest in geology and enthusiasm for collecting. He was acknowledged as the local expert on gemstones, which was why George Milner Stephen sought him out during his visits to Beechworth in the early 1860s. In 1864, the Geological Survey of Victoria invited Dunn to join its staff and his first posting was to George Ulrich's geological mapping camp near Maldon. The meeting was fortuitous, as Ulrich was an inspirational teacher and Dunn an eager pupil. After a fruitful four years of geological mapping, Dunn was cast adrift when the Survey was disbanded in 1868, just prior to his qualifying as a mining surveyor. He returned to Beechworth but, frustrated by inactivity, successfully applied for a position as Government Geologist in Cape Colony, South Africa, in 1871. There he embarked on pioneering surveys of the rich diamond, gold and coal deposits, collecting rocks, minerals and tribal artefacts on his travels. His skills as a field geologist came to the fore and he made several important discoveries. In 1883 he predicted the enormous richness of the Transvaal gold deposits. Dunn had married in 1875 during one of several visits to England, and in 1886 decided to return with his family to Victoria, where he took up a consulting career. This saw him travelling widely throughout

GEORGE HENRY FREDERICK ULRICH

Born in Zellerfeld, Germany, in 1830, George Ulrich was educated at the Royal Prussian School of Mines in Clausthal, graduating in 1851. He arrived in Victoria in August 1853 and spent nearly three years as a digger on the goldfields. In September 1857 he joined the Geological Survey, at that time engaged in systematic mapping of the goldfields. Ulrich's skills enabled him to reach the rank of Senior Field Geologist, then to embark on a career as a consulting geologist after the Survey was disbanded in 1868. In June 1870, he was appointed Curator of the Mineral Collections and Lecturer in Mineralogy in the newly-founded Industrial and Technological Museum in Melbourne. His work at the Museum continued until March 1878, when he moved to New Zealand following his appointment as Professor of Mining and Mineralogy at the University of Otago and Director of the Otago School of Mines in Dunedin. Ulrich's distinguished career came to a tragic end in 1900, when he was fatally injured in a fall on a collecting trip.

Australia, visiting and reporting on many mines, including a landmark study of the Bendigo goldfield. In 1904 he was appointed Director of the Geological Survey of Victoria, and continued to travel and publish reports, before retiring in 1912. During his career Dunn wrote three books (*Pebbles*, *Geology of Gold* and *The Bushman*), and published over 200 reports. Perhaps equally importantly, he was a discerning collector who donated significant geological and ethnographic material to museums throughout his lifetime. The most significant portion of his collection was acquired by the National Museum of Victoria in 1948, some years after his death in Melbourne in 1937. Many old specimens from Victorian and South African mines bear his hand-written labels. The most treasured items, however, are the gemstones Dunn collected during his boyhood wanderings around Beechworth.

The intense mining activities in Victoria between the 1850s and 1870s resulted in a great many new mineral finds, a situation in which Ulrich revelled. He visited most of the mining areas and mineral localities in the State and published articles describing many of his discoveries. Perhaps his best-known works are his essay *Mineral Species of Victoria*, written for the Intercolonial Exhibition held in Melbourne in 1866–7, and his *Contributions to the Mineralogy of Victoria* in 1870. The latter contains his description of the new species maldonite. As Victoria's only classically trained mineralogist, George Ulrich's knowledge of and enthusiasm for the State's mineralogy was unsurpassed. His contributions were recognised by the naming of another new Victoria mineral, ulrichite, in 1988.

Geological setting

The wide range of gem minerals recorded in Victoria is a reflection of the region's geological diversity. For much of the past 600 million years, southeastern Australia has been close to active zones between large plates of the Earth's crust. This has seen chains of volcanic islands form, only to be destroyed by forces which opened and closed ocean basins several times. Layers of sedimentary rocks kilometres thick have been folded, faulted or metamorphosed by these same forces, then intruded by granite magmas. Mountain ranges have been thrust up and worn down by erosion. Parts of the region have been blanketed by layers of ash from huge volcanic eruptions, and scoured by glaciers and ice-sheets. Shallow seas have invaded the land, then retreated. Long-established river systems have been disrupted and rearranged by earth movements, or buried by lava flows. All these processes created environments for the formation of a wide range of rocks and minerals, including those used as gems or ornamental materials.

Most of the gemstones are associated with two major groups of rocks. These are the granites which were intruded during the Silurian and Devonian periods, and the volcanic rocks which erupted throughout the Cenozoic era. Minerals such as topaz, tourmaline and some coloured varieties of quartz, with less common beryl and garnet, crystallised late in the cooling history of many of the granite intrusions. By the time the basaltic lavas of the so-called Older Volcanics were erupted, between about 65 and 5 million years ago, erosion had exposed the granites and their small scattered gem mineral occurrences. As the rocks weathered, crystals were released into the soils and eventually found their way into local streams. There they were often joined by sapphires, rubies and zircons, possibly even diamonds, weathered out of the volcanic rocks. A feature of Victorian gem deposits is that many crystals have been through several cycles of erosion and deposition and have been carried long distances before ending up in modern-day streams. Erosion also released and redeposited grains and lumps of gold from quartz veins scattered through the folded sedimentary rocks across Victoria, so that gemstones and gold frequently coexist in the stream gravels.

Other geological environments in Victoria have produced fewer and less valuable gemstones. The basalt flows of the Western Victorian volcanic plains are less than 5 million years old, and in some places yield olivine in its gem-quality variety, known as peridot. The oldest rocks in the State, altered volcanic rocks erupted in the Cambrian period, yield few gem minerals, although rocks such as coloured chert and jasper, and rare corundum, occur with them. The glaciers of the Permian period moved generally northwards across the landscape, at a time when Australia and Antarctica were still joined. They brought with them a wide range of foreign rocks and minerals, which were dumped when eventually the ice melted and retreated. Exotic pebbles of granite, rhyolite, agate and corundum may have come from sources which now lie buried beneath the Antarctic icecap.

The types of geological processes which form gem minerals are not unique to Victoria. They have affected not just our region but all of the continents as we know them throughout geological time. The familiar rocks and minerals which result from these global processes enable geologists to make comparisons and predictions, just as George Ulrich did when he first examined the parcel of gemstones from Guildford back in 1860.

1.7 *(opposite)* Geological environments for Victorian gemstones.

TIMING OF GEMSTONE FORMATION IN VICTORIAN GEOLOGICAL HISTORY

ERA	PERIOD	EPOCH	AGE*	GEMSTONES & ENVIRONMENTS
CENOZOIC	Quaternary	Holocene	0.01	Gemstones and pebbles in present day streams.
		Pleistocene	2.6	
	Neogene	Pliocene	5.3	Basaltic lavas provide olivine and anorthoclase. Sapphires and zircons in Pliocene stream gravels.
		Miocene	23	
	Paleogene	Oligocene	34	Silicified wood.
		Eocene	56	Volcanoes erupted at intervals from Paleocene to Miocene, and provide sapphires, zircons, olivine and possibly diamonds.
		Paleocene	66	
MESOZOIC	Cretaceous		145	Polished pebbles.
	Jurassic		201	Pyrope garnets in volcanic pipes.
	Triassic		252	
PALEOZOIC	Permian		299	Glaciers and ice-sheets bring in foreign rocks and minerals such as agates and corundum.
	Carboniferous		359	
	Devonian		419	Quartz crystals in reefs. Agate and chalcedony in geodes in acid volcanic rocks. Granite intrusions with quartz crystals, tourmaline, topaz, garnets, beryl, cassiterite.
	Silurian		444	
	Ordovician		485	Quartz crystals in reefs. Slates are host rocks for turquoise.
	Cambrian		541	Chert, jasper, 'selwynite' hosted by greenstones (altered volcanic rocks). Garnets.

*Millions of years ago

Diamonds

"I may remark that all the Beechworth diamonds that I have seen — about a dozen — were beautifully distinct in their crystallographic features."

REV. JOHN BLEASDALE, 1863

◇◇◇◇◇◇◇◇◇◇◇◇◇◇◇◇◇◇◇

Small diamonds have been found in alluvial deposits in Victoria, from the northeast to the far west. The main sources are gold-bearing gravels in present-day stream valleys and older buried stream channels, known as deep leads, in the Beechworth–Eldorado and Chiltern–Rutherglen areas. These regions were part of the Ovens District mining field, centred on Beechworth. Most of the diamonds were discovered by accident during treatment of wash-dirt for gold and tin recovery between the 1860s and early 1900s. Despite investigations by mining companies, no commercial deposits have been found and the primary source of the diamonds remains unknown.

DIAMOND
(Beechworth)

2.1 Diamond crystal sketches,
Beechworth (Ulrich, 1866).

2.2 *(opposite)* Woolshed Goldfield in 1858,
by Alfred Eustace (43 cm x 30 cm)
(Burke Museum collection, Beechworth).

History of discovery

REPORTS OF FINDS

Victorian diamonds were first reported in the early 1860s. Alfred Selwyn, Director of the Geological Survey, made the earliest official record of their discovery. In his 1860 Annual Report (dated May 1861), Selwyn added diamonds from the gold drifts to the 'commercially valuable mineral products of the State'. Selwyn may have been influenced by the tabling of a small Ovens District diamond in the Legislative Assembly by George Stephen, then the Member for Collingwood, in May 1860. While the resulting parliamentary discussion was frivolous, it prompted a response from the noted natural historian Ludwig Becker, who wrote to the *Argus* newspaper (12 May) challenging the stone's authenticity and declaring it to be a topaz. This identification was immediately refuted by local lapidary James Spink (*Argus*, 14 May), and Becker was forced into a retraction, in which he agreed that, after further examination, the stone was a diamond after all (*Argus*, 15 May).

In the *Herald* of 28 July, 1860, George Ulrich referred to reports of a diamond found in gravels at Yandoit, near Daylesford. This is probably the small colourless diamond reported from Sandy Creek, near Jim Crow Diggings (Daylesford), in *The Argus* of 25 June 1860. At the Victorian Exhibition held in Melbourne in 1861, several small 'not fully verified' diamonds from the Ovens goldfields were displayed. A diamond found at Yackandandah was exhibited at the International Exhibition in London in 1862. Probably the first description of a Victorian diamond appeared in *Chemical News*, 5 July, 1862, and was based on an examination of a stone from the Ovens District by Mr George Foord. The diamond weighed 2.356 grains (0.74 carats) and was a pale-yellow, modified dodecahedron with

Wool Shed 1858

diamond finds in the Beechworth area. In the Royal Society's 1865 exhibition of colonial gemstones, ten Ovens District diamonds were on display, including one uncut stone mounted in a ring. The first illustrations of a Victorian diamond were sketches of crystals from Beechworth accompanying George Ulrich's essay, *Mineral Species of Victoria*, in 1866.

From 1864, official records on diamond finds were kept by the Mining Warden in Beechworth, and published in the Annual Mining Statistics of Victoria (Smyth, 1865–1875; Couchman, 1876–1883). Up until 1875, 102 diamonds were reported from the creeks in the Ovens District, most from the Woolshed Valley upstream from the small town of Eldorado. From this time on, the number of alluvial claims working the creeks for gold and tin declined rapidly. As a result, diamond discoveries stopped, with the last report being of a small stone found in 1879.

These official records are unlikely to represent all the diamonds discovered during early alluvial mining in the Beechworth district. In 1867, Alfred Selwyn suggested that because diamonds were only recognisable to the experienced eye, and that those recovered were only found during gold panning after all pebbles greater than $1/4$ inch (6 mm) had been removed by sieve, 'it is not improbable that not only a number of small but more especially the larger, and therefore far more valuable, gems are annually overlooked'. Based on his observations as a resident of Beechworth during the 1860s, Edward Dunn estimated that between 200 and 300 diamonds were recovered during treatment of the wash-dirt (Dunn, 1913).

Between 1869 and 1873, the Mining Registrar at Sandhurst (Bendigo) reported several diamonds being found at Huntly and other localities in

COLONIAL GEMS—UNPOLISHED.

DISGUSTED DIGGER *(to Mate).*— *Well*, TOM, *if those are diamonds, strikes me we've thrown away a fortin' in our time.*

◊ **2.3** 'At the Royal Society's Exhibition'. Cartoon from *Melbourne Punch*, May 4 1865.

convex striated faces. At a meeting of the Royal Society of Victoria in November 1863, Reverend John Bleasdale exhibited three diamonds. Two were from Beechworth, and the third was a small diamond said to have been found at Collingwood Flat, Melbourne, on a garden path made from gravel quarried from nearby Northcote or Abbotsford. In 1863 and 1864, articles in the *Argus* and *Dickers Mining Record* (Baragwanath, 1948) described various the district, but there are no means of authenticating these discoveries. The *Argus* of 27 April, 1895 reported that Melbourne jewellers had authenticated diamonds found in Pliocene drift near Congbool (now Kongbool) in far western Victoria. The Government Geologist, James Stirling, visited the locality and described one of the stones (Stirling, 1898b). One of several diamonds found in 1896 in wash-dirt in the remote Toombullup goldfield was sent to Melbourne, where the Secretary for Mines had it examined and verified.

Diamonds made a reappearance in the Ovens District during the working of the Great Southern gold mine at Rutherglen between 1909 and 1914. About 20 small stones, the largest about 1.3 carats, appear to have been recovered, most by a Mr James Graham (Baragwanath, 1948). As at Beechworth, these diamonds were found by accident. In 1912, the District Mining Inspector suggested the company put in more sluice-boxes with greased blankets to save any diamonds, but there is no indication of any further discoveries. In 1947, Graham sold nine Great Southern diamonds to the National Museum of Victoria for £15.

Since the Rutherglen discoveries, there have been only sporadic findings, but these include the two largest diamonds found in Victoria. Late in 1916, a 6-carat stone was found in a patch of wash-dirt at the junction of Black Sand and Old Hands

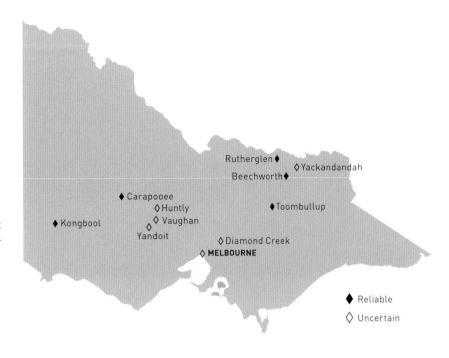

creeks, near Beechworth, by a Mr D. Methven (Baragwanath, 1948). In 1977, a prospector using a greased shovel unearthed the largest stone so far recovered in Victoria, weighing 8.2 carats, in the gravel of Black Sand Creek. Diamonds were supposedly found regularly during dredging operations in the valley of Reids Creek, near Eldorado, between 1934 and 1954, but no records were kept. There are still occasional reports of small diamonds being found by fossickers in the Woolshed Valley. Late in 1993, a 0.5-carat, pale-yellow diamond was found on a vibrating table in the processing plant of Southern Border Gold Ltd's alluvial gold mine at Carapooee, 12 km SSE of St Arnaud. In 2008,

2.4 Main locations where diamonds have been reported in Victoria.

fossickers found a very small diamond in gravels panned from a tributary of Middle Creek, in the Toombullup district. Their discovery lends credibility to the reports of diamonds from this goldfield in 1896.

Records of diamonds from Fifteen Mile Creek (Herman, 1914) and Mansfield (Chalmers, 1967) almost certainly refer to the Toombullup occurrence. Small diamond crystals, possibly from the Ararat–Stawell region, were described by Poth (1984a). Reported diamonds at Lake Bullenmerri, near Camperdown, were based on contaminated samples.

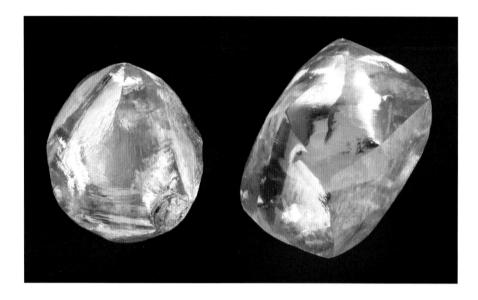

USE OF THE DIAMONDS

At the time of the early discoveries, Victoria's diamonds were of sufficient scientific interest to be featured in the Colonial and International Exhibitions held throughout the 1860s. Their faceting qualities generated mixed opinions, although some of the Beechworth diamonds were purchased by jewellers, such as George Crisp in Melbourne and William Turner in Beechworth, at prices between £2 and £3 per carat. One of the Beechworth diamonds exhibited by Bleasdale in 1863 weighed 3 carats before it was sent to Amsterdam, where it was transformed into a 'magnificent gem', weighing 2 carats and with a value between £35 and £40. A brilliant cut, it was mounted as a brooch and displayed by James Murray in the Royal Society's exhibition in 1865. A selection of diamonds from the Great Southern mine at Rutherglen was sent to Amsterdam for cutting before being presented to directors of the company.

The whereabouts of any of these early cut Victorian diamonds are unknown, but the scientific interest in the diamonds persists to the present day.

2.5 *(top)* Diamonds from the 'Ovens diggings' (Natural History Museum collection, London; smaller is 4 mm across).

2.6 Diamonds from the Ovens district formerly in E. J. Dunn's collection (largest is 0.95 carats, 5.5 mm across).

SURVIVING DIAMONDS

There are about 30 Victorian diamonds known
to be held in museums or private collections.
Museum Victoria has eight from the Beechworth–
Eldorado area, eight from Rutherglen, one
from Carapooee, one from the Toombullup
district and two reported to be from Diamond
Creek. One of the nine Rutherglen diamonds
bought from James Graham has been lost
and the others do not correspond with those
listed by Baragwanath (1948) as being in
Graham's possession. The Burke Museum
at Beechworth has five local diamonds in
its collection. There is an Eldorado diamond
in the collection of the Geology Department
at the University of Melbourne. The Natural
History Museum (London) has two Beechworth
diamonds, including one of 0.4 carats purchased
from Alfred Selwyn in 1864. Those diamonds
presented to the former Mines Department
(Geological Museum) by Edward Dunn (see list
in Baragwanath, 1948) were transferred to the
Museum of Victoria in 1987 (one of these, from
the Lancashire Lead, near Chiltern, had been
lost after an exhibition in 1979). The 8.2-carat
Beechworth diamond is still in the possession
of the finder. Two diamond crystals found by a
worker on the Cocks Eldorado Dredge in the
early 1930s were owned by a relative of the finder.
One stone, weighing 0.93 carats, was purchased
by CRA Exploration Pty Ltd during the early
1980s, but was destroyed during examination.

Two more Eldorado diamonds, weighing 1.6 carats
and 1 carat, have been acquired by Museum
Victoria from private collections since 2000.

2.7 Diamond crystal (1 carat) from Eldorado
(Private collection).

RECENT EXPLORATION

Volcanic pipes with similarities to kimberlite —
the best known but not the only host rock for
diamonds elsewhere in the world — have been
the traditional exploration targets for Victorian
(and New South Wales) diamonds. However,
because of their small size (perhaps less than
100 m across) and highly weathered nature,
such pipes are probably as difficult to find as
the diamonds themselves, especially if they
are covered by younger sedimentary deposits.
A weathered pipe near Meredith, about 50 km
west of Melbourne, has been investigated
in detail (Day *et al*, 1979), but it is not in a
region where diamonds have been found.
CRA Exploration Pty Ltd has used geophysical
methods to find small targets buried beneath the
younger sedimentary rocks of the Murray Basin
in northwest Victoria. Drill cores show that some
of these resemble kimberlite, but no diamonds
have been reported. A number of companies,
including Ashton Mining Ltd, have explored
for diamonds in central and northeastern
Victoria during the past two decades, but no
significant discoveries have eventuated.

Mining companies conducting regional exploration programs in Victoria, especially in the northeast, have focused on so-called 'indicator minerals'. These are minerals which are typically found with diamonds in kimberlite. They have densities in the range 3.25 to 4.6 g/cm^3 and concentrate in the heaviest fraction in alluvial deposits. The main indicator minerals are pyrope (the magnesium-rich member of the garnet group), magnesium-bearing ilmenite ('picroilmenite'), and spinel and diopside containing significant amounts of chromium. While these minerals do occur in small amounts in Victoria, they are uncommon or absent in the regions where diamonds have been recorded.

2.8 Woolshed Falls, Reedy Creek (1996).

Localities

OVENS DISTRICT

Two major deep lead systems occur in the Ovens District, about 240 km northeast of Melbourne. These contained rich deposits of gold and tin, as well as being the main source of the State's diamonds. The Eldorado–Beechworth system in the south and the Chiltern–Rutherglen system in the north are separated by the main divide of the Mount Pilot Range, but share similar mining histories and geological features. Beechworth is the major settlement in the district and is one of Victoria's most historic towns. About 7 km northwest of Beechworth is the picturesque Woolshed Valley, the best known of Victoria's gemfields. It forms part of the main Eldorado Lead system (Canavan, 1988), which extends along the valley of Reedy Creek

from a few kilometres west of Eldorado to the headwaters of its two main tributaries, the SSW-flowing Wooragee Creek and the NNW-flowing Spring Creek. For much of its length, Reedy Creek has cut deeply into granite bedrock, forming rapids and waterfalls, but the valley is more open near Eldorado, at the Woolshed and near Wooragee. North of Beechworth, Spring Creek flows through a narrow gorge cut in granite. Mining for gold or tin took place intermittently over the 35km length of the Eldorado Lead system between 1853 and the early 1950s.

The Chiltern–Rutherglen region forms part of the flood-plains of the Murray and Ovens rivers. The main present-day streams drain

off the granite forming the Mount Pilot Range in the south, joining to form Black Dog Creek, which flows NNW into the Ovens River. Buried beneath the wide alluvial plains lies a complex network of former river and stream deposits. The courses of the main deep leads twist and turn for nearly 50 km, from SE to NW, and are joined by many shallower tributaries. This system was exploited by the most extensive development of deep lead mining in the State from the 1860s to the early 1900s.

History of mining
The land around the present town of Beechworth was taken up in 1838 by the Reid brothers, who established a sheep station at Wooragee. Gold was discovered by a group of three men in

2.9 View over Chiltern-Rutherglen region, looking north from Mt Pilot.

nearby Spring Creek early in 1852. A small rush started, and the original finders moved downstream, opening up rich deposits in Reids Creek (Flett, 1970). There was a brief exodus from the Beechworth field when rich gold was discovered at Bendigo, but diggers gradually returned and by late 1852 there were about 1500 men working Spring and Reids creeks. Woolshed Creek was opened up in 1853 and by 1855 was the richest goldfield in the area. Even further downstream, diggings at Sebastopol and Eldorado were commenced in 1855, while upstream, the Wooragee field was first worked in 1856. From about 1854, alluvial cassiterite —

◇

2.10 Cocks Eldorado dredge (1996).

2.11 *(opposite)* Simplified geological map of the Beechworth–Chiltern–Eldorado district.

tin oxide — was recovered with the gold in such quantities that the valley became the largest tinfield in the State. The gold ran out by the early 1860s, but tin mining by large-scale dredging continued in the western part of the valley until 1954.

Gold discoveries in the Chiltern–Rutherglen region were a little later than those in the Woolshed Valley. The Indigo Lead, about 8 km north of Chiltern, was the first to be prospected. A major rush took place late in 1858, and by 1859 there were 13 000 diggers on the field. Other rich shallow leads, such as the Wahgunyah diggings, were opened up in the region between 1859 and 1861. As the rich shallow deposits were exhausted, interest turned to the deeper buried leads. The Lancashire Lead, forming the southeastern end of the system, was opened in about 1867. Gradually the system was traced downstream, although not always mined, by a combination of bores, then shafts and underground workings, until the main Chiltern Valley Lead was

reached, a few kilometres west of Chiltern. Companies in this 6.5 km stretch of the lead system operated from 1870 to 1920 and produced about 10 000 kg of gold (Canavan, 1988). Further north, nearer Rutherglen, the Prentice Lead was opened up around 1888 and companies such as Great Southern Gold Mining operated until about 1916, producing about 31 600 kg of gold. After the 1920s, mining of the deep leads ceased due to lower gold grades and large inflows of water. Several mining companies investigated the area for gold and tin during the 1960s, but no further production has occurred.

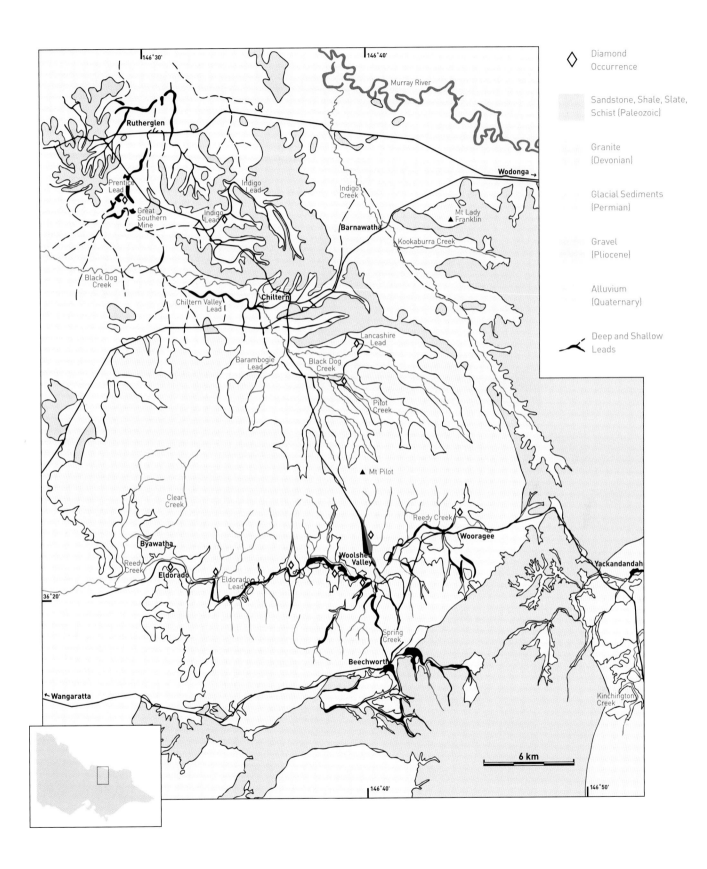

Diamond
Occurrence

Sandstone, Shale, Slate,
Schist (Paleozoic)

Granite
(Devonian)

Glacial Sediments
(Permian)

Gravel
(Pliocene)

Alluvium
(Quaternary)

Deep and Shallow
Leads

146°30'

146°40'

Murray River

Rutherglen

Wodonga →

Prentice
Lead

Indigo
Lead

Indigo
Creek

Mt Lady
Franklin

Great
Southern
Mine

Indigo
Lead

Barnawatha

Kookaburra Creek

Black Dog
Creek

Chiltern Valley
Lead

Chiltern

Lancashire
Lead

Barambogie
Lead

Black Dog
Creek

Pilot
Creek

Mt Pilot

Clear
Creek

Reedy Creek

Byawatha

Wooragee

Reedy
Creek

Woolshed
Valley

Yackandandah

Eldorado

Eldorado
Lead

36°20'

Spring
Creek

Beechworth

← Wangaratta

Kinchington
Creek

6 km

146°40'

146°50'

2.12 Diamond crystal from the Ovens district (0.2 carats, 3 mm) (Burke Museum collection, Beechworth).

2.13 Beechworth diamond (2 carats) (University of Melbourne collection).

Geology and mineralisation

Two groups of rock types form the bedrock of the Ovens District. Low-grade, regionally metamorphosed Upper Ordovician sedimentary rocks have been intruded by a large complex granite pluton, of Late Devonian age. The main mass is the Beechworth Granite, but a number of smaller related intrusions have been identified (Rossiter, 2003; Young, 1983). The southern margin of the granite trends roughly NE–SW through Beechworth, and the northern margin runs E–W about 2 km south of Chiltern. Overlying the granite, several areas of presumed Permian glacial pebble conglomerates and

tillites have been preserved, especially near Wooragee in the east and Eldorado in the west.

Extensive erosion of the granite and metasediments has led to the former drainage system being filled by thick alluvial deposits, ranging in age from the Early Pliocene to the present day. These sedimentary rocks include the deep lead systems. In the Chiltern–Rutherglen region, the alluvial deposits are between 75 and 120 m thick in the main valleys. The Eldorado Lead system, in the Woolshed Valley, is mostly in shallow Recent stream gravels less than 12 m thick. However, at the western end of the valley near Eldorado, the alluvium is between 70 and 80 m thick. Older Pliocene alluvial deposits form cemented cappings on lower spurs in parts of the Woolshed Valley.

The distribution of gold in the deep and shallow lead deposits indicates that it was almost all derived from quartz reefs in the Ordovician sediments. Over 300 reefs were worked in the ranges southeast of Beechworth, and probably a similar number in the low hills northwest of Chiltern. Black cassiterite-bearing sand occurred in every creek over the entire granite area and was derived from small lodes scattered within the Mount Pilot Granite. Other minerals found in small quartz veins in the granite include molybdenite, scheelite, loellingite and bismuth.

From the earliest mining operations in the Woolshed Valley, the beautiful coloured gem minerals found with the gold after panning of the wash-dirt attracted attention. Each dish showed topaz, blue, green and yellow sapphires, black tourmaline and spinel, and yellow to orange–red garnet and zircon. Occasionally a diamond was found. Similar gem minerals were present in several of the Chiltern–Rutherglen deep leads running off the granite area, for

example, the Lancashire Lead (Dunn, 1871) and the Chiltern Valley Lead (Dunn, 1887, 1909a).

Such a varied assortment of gem minerals must have come from different source rocks over a long period of time, with the Woolshed Valley resembling a giant sluice box. The granite provided topaz, tourmaline, possibly some garnet and several varieties of coloured quartz, as well as the cassiterite. The Permian glacial conglomerates provided rounded pebbles of jasper, chert and agate, as well as opaque corundum of various colours. Sapphire, zircon and black spinel (pleonaste) were probably derived from basic volcanic rocks which have now been eroded away.

Distribution of the diamonds

Most of the diamonds in the Eldorado Lead system have been found in alluvial gravels in Reedy Creek — in its sections known earlier as Wooragee Creek, Youngs Creek and Woolshed Creek — from Wooragee to Eldorado. Of the 78 diamonds for which more precise locality details other than 'Beechworth' are recorded, 34 were found at the Woolshed, nine at Sebastopol, nine at Eldorado, six at Youngs Creek, five at Wooragee, five at Napoleon's Flat and one at Sheep Station Creek. Several of the Sebastopol diamonds were found in the cemented Pliocene gravels (Dunn, 1871). The greatest number from one claim was 15, obtained at Finns claim on the Woolshed.

2.14 Diamonds from the Ovens district (largest is 1.2 carats, 6 mm across) (Burke Museum collection, Beechworth).

2.15 (*above*) Diamonds from Eldorado (larger is 1.6 carats, 5 mm across).

2.16 (*below*) Diamonds from the Great Southern mine, Rutherglen (largest is 0.8 carats, 4.5 mm across).

Ten stones were obtained in a 30 x 15 ft (4.5 x 9 m) cut by taking an occasional dish from the sluices and washing it carefully (Dunn, 1871). At Eldorado, the western limit of the diamond discoveries, the stones were found in the bottom level of the alluvial deposits, on granite bedrock. Black Sand Creek (also known as Snakes Head Creek) and Old Hands Creek have supplied the two largest stones; the 8.2-carat diamond found in a metre-deep hole in Black Sand Creek upstream from the present bridge, and the 6-carat stone at the bottom of a wash deposit in 1917 at the junction of the two creeks. No diamonds have been found in Spring Creek.

Diamonds were found in several leads in the southeastern headwaters of the Chiltern–Rutherglen system. According to Reports of the Mining Registrar at Beechworth, four diamonds were found at Pilot Creek between 1872 and 1875. Where along Pilot Creek these were discovered is unknown, as there are no production records for this lead, but the remains of shallow diggings can be seen in the headwaters of the creek. It is known that, after 1867, wash-dirt was proved in bores, followed by the sinking of several shafts, in parts of the Lancashire Lead close to the Pilot Tin and Gold Mining Company's shaft (Hunter, 1909). Dunn (1871, 1909a) listed the Lancashire Lead as a source of diamonds; in 1887 he referred to a small diamond he had collected in this lead 'several years since' (this is probably the diamond lost from the Geological Survey collections in 1979). Dunn (1871) also listed diamonds occurring at the Indigo Lead, but provided no details.

The largest concentration of diamonds in the Chiltern–Rutherglen field came from the Great Southern Gold Mining Company, which had leases on the Prentice Lead about 3 km southeast of Rutherglen. The diamonds were found between 1911 and 1913 at a depth of 115 m (350 ft). Despite the extensive working of the black cassiterite-bearing sand in the leads upstream and downstream of the Great Southern gold mine, no other diamonds were reported in the Prentice Lead. Hunt (1978) presented evidence which suggested these diamonds came from a patch of wash in the upstream portion of the lead, near the lease boundary with the Great Southern Consols mine.

Features of the diamonds

From published descriptions and observation of the few stones remaining in collections, it is possible to summarise the main features of the Ovens District diamonds. However, complete gemmological and mineralogical studies remain to be done.

Weight and dimensions

Most of the diamonds from the Eldorado lead system weighed less than 0.5 carats, with about 20 recorded as being between 1 and 2 carats. Single stones of 3, 5, 6 and 8.2 carats have been found. A Mining Surveyor's report of a 17.64-carat stone found at Sebastopol in December 1864 was corrected in 1866 — to read 1.764 carats. Apart from the two largest diamonds from the Black Sand Creek gravels, the weights of the remaining stones do not appear to show any significant relationship with distribution in the lead system. The 8.2-carat diamond from Black Sand Creek measures 12 x 10.5 x 10 mm. The maximum dimension of any of the other available diamonds is about 7 mm.

2.17 Distribution of reported diamonds in the Eldorado Lead system (historical creek names).

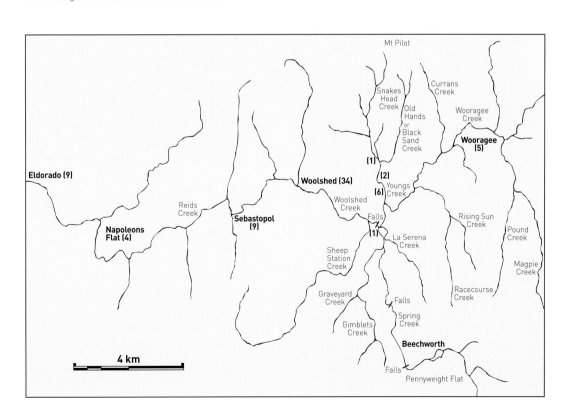

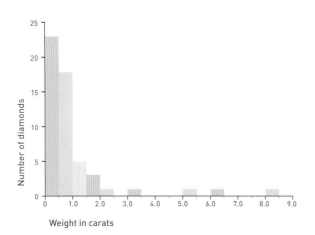

◊ **2.18** Weight distribution of Oven's district diamonds.

2.19 Diamond from Black Sand Creek (8.2 carats, 12 mm across) collected in 1977 (Private collection).

In the Chiltern–Rutherglen lead system (which includes the Pilot Creek lead), the recorded diamonds range from less than 0.03 carats up to 1.5 carats. The maximum dimension of any stone is 7 mm.

Shape and twinning

Many of the diamonds from the Eldorado lead system have been described as variations on the octahedron. These are mainly trisoctahedrons which are often elongated or flattened, or modified by additional small faces. Simple forms, such as those illustrated by Ulrich, were the exception. Instead, most crystals show convex faces with curved edges, giving them a rounded appearance. Some rounded trisoctahedrons closely resemble dodecahedrons and tetrahexahedrons. For example, the five diamonds formerly owned by Dunn were described as dodecahedrons by Hall and Smith (1982). However, close examination shows the 'rhombic' faces on at least three crystals are divided into two unequal triangular faces by an indistinct edge, so the crystals are distorted tetrahexahedrons. None of the Eldorado Lead diamonds show twinning.

The Great Southern mine diamonds range from nearly spherical to slightly flattened stones which appear to be mainly rounded tetrahexahedral forms, many showing a faint twin plane. Several irregular stones are present.

Colour and lustre

The Eldorado lead diamonds range from colourless to pale and straw yellow. Pale bluish and brownish tinges are present in some near-colourless stones. Dunn (1871) reported that a few diamonds were green or pink, but none showing these colours has been preserved. The lustre is quite variable. Most of the yellow stones have bright surfaces, but with faint frosting giving them a slightly greasy lustre. Several of the colourless and pale-yellow stones are quite dull. The Great Southern mine stones are colourless to pale yellow and have a bright lustre.

Abrasion features

Abrasion of edges and apices of the diamonds ranges from slight for the Pilot Creek diamond, to mild in most of the Eldorado Lead stones, and significant for some of the Great Southern stones. Many of the diamonds show graphite-free fractures penetrating from the flat faces, as well as percussion marks. These features may be due to high energy impact, either during alluvial transport or possibly during eruption.

Other surface features

Surfaces of some of the Eldorado Lead diamonds were described as showing prominent 'roll-like' relief by Hall and Smith (1982). Faint surface cracks are also present. Prominent dissolution lamellae which reveal strong zoning occur on several of the diamonds, and growth triads are also present on a few stones.

Many of the diamonds from the Great Southern gold mine show pits and shallow euhedral indentations up to 1 mm across representing former crystals, possibly of garnet or pyroxene. Such impressions on Eldorado Lead stones are smaller and less common.

Inclusions

Many of the Eldorado diamonds show small near-surface green or brown spots (Hall and Smith, 1982). These are due to radiation damage, brought about by close, fixed contact with grains of radioactive minerals, such as zircon and monazite, over a prolonged period. Green spots turn brown with heating, such as from lava flows or bushfires. The presence of green and brown spots on the Pilot Creek diamond suggests two periods of fixation, separated by a heating episode. In the Eldorado Lead and Great Southern mine stones, there are possible coesite inclusions and some internal fractures lined with graphite. A chromite inclusion was identified in the Eldorado diamond purchased by CRA Exploration.

Composition

The minor element contents of Victorian diamonds have not yet been determined. However, the pale to moderate yellowish colours of any of the Beechworth diamonds suggest nitrogen contents between 500 and 1000 parts per million (Taylor *et al.*, 1990).

The source of the diamonds

The most reliably located diamonds in the Eldorado and Chiltern–Rutherglen lead systems all occur in deep leads or stream gravels which

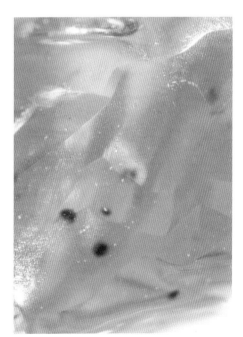

2.20 Radiation spots (green and brown) on diamond from Pilot Creek, Beechworth district.

have drained off the main divide of the Mount Pilot Range, in granite bedrock. The simplest explanation for this distribution, assuming the diamonds are younger than the Permian glacial deposits, would be a source or sources, such as one or more volcanic pipes, on the divide east of Mount Pilot. This would allow south-draining streams such as Old Hand (Black Sand) and Snake Head creeks, and unnamed tributaries entering Reedy Creek near Wooragee, to transport diamonds into the Eldorado Lead system. Streams flowing northwards off the divide, such as Pilot Creek (in which diamonds were found) and the headwaters of Black Dog Creek, could have carried diamonds into the Lancashire Lead and, possibly, as far downstream as the Great Southern gold mine.

However, there are some differences between the features shown by the Eldorado and Great Southern mine diamonds, suggesting they may have different origins and histories. The more abraded nature of the Great Southern diamonds and their lack of radiation spots suggest they have undergone high-energy transport and were then enclosed in alluvial deposits with low zircon concentrations. The radiation spots in the Eldorado Lead and Pilot Creek diamonds could be explained if the stones had been enclosed in the older cemented Pliocene deposits, before being released into Recent alluvial gravels. The report by Dunn (1871) that some diamonds had been found in these older cemented gravels at the Woolshed, and that one Wooragee diamond had a film of iron oxide, would support this suggestion.

Dunn himself believed the diamonds were not of local origin but were being shed from the Permian glacial conglomerates (Dunn, 1913). These deposits would originally have been more widespread than at present, allowing distribution of diamonds into the deep leads from early Pliocene times. In this case, the differences in abrasion and radiation spots between the Eldorado and Great Southern mine diamonds could reflect local differences in transport and deposition conditions during the early Pliocene. Dunn suggested that volcanic pipes similar to those near Delegate, in southern New South Wales, were a possible source rock. However, these volcanic rocks are now known to be Mesozoic in age and hence could not have supplied diamonds for the older glacial deposits.

The Beechworth–Chiltern area, especially the Wooragee valley, has been the focus of intermittent diamond exploration over the last 40 years. A combination of heavy mineral analysis to find indicator minerals and drilling of geophysical anomalies has been used by companies including Northern Mining Corporation NL, Penzoil of Australia Ltd, Freeport of Australia, Triad Minerals NL, Poseidon Exploration, Warren Jay Holdings Pty Ltd, and Stellar Resources Ltd/Rimfire Pacific Mining NL. CRA Exploration's examination of Eldorado Lead concentrates in the Museum of Victoria's collections in the early 1980s identified ilmenite with basalt–kimberlite affinities and some chromite typical of kimberlitic rocks, but no garnets rich in magnesium or chromium. Chromite and chromium-bearing diopside in concentrates from Magpie Creek near Wooragee were suggested by Tangney (1991) to have a possible kimberlite source in nearby granite covered by Pliocene sediments.

The results of the exploration programs in the Wooragee valley have been inconclusive and subject to varying interpretations, and no primary source has been located. The age and origin of the Ovens District diamonds remain a mystery to this day.

TOOMBULLUP

The Toombullup goldfield is small and isolated, situated about 40 km southeast of Benalla or 25 km northeast of Mansfield. Gold was first found in the area in 1858 (Flett, 1970), but the main rushes were not until the late 1890s. The diggings were in irregular patches of auriferous Cenozoic clays, gravels and conglomerates on a basement of Devonian acid volcanic rocks (originally referred to as granite), near the western headwaters of Ryans Creek (Stirling, 1898a). This area, in the upper reaches of present-day Dogwood Creek, a tributary of Ryans Creek, was called Puzzle Gully by Kitson (1896). Gold was also obtained from younger leads in nearby streams such as Stringybark, Webbs and Middle creeks, and gemstones such as sapphires and zircons occurred with the gold at Puzzle Gully, Webbs Creek and Middle Creek (Kitson, 1896). Remnant patches of Cenozoic basalt cap low hills in the region, and have ages between 36 and 43 million years (Wellman, 1974). Sub-basaltic gravels and coal seams near Toombullup contain leaf fossils of probable Eocene age (Howitt, 1906b).

2.21 Simplified geological map of Toombullup district (note section line for Fig. 2.22).

2.22 (*below*) Sketch cross section (from Stirling, 1898a) (see map 2.21).

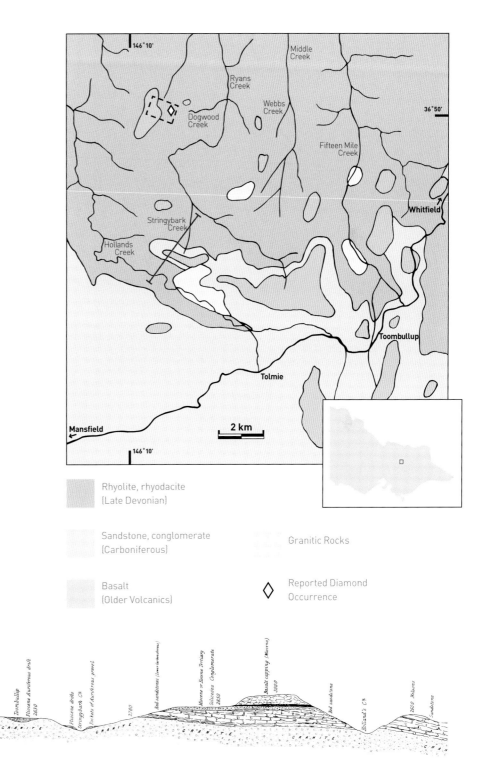

Rhyolite, rhyodacite
(Late Devonian)

Sandstone, conglomerate
(Carboniferous)

Granitic Rocks

Basalt
(Older Volcanics)

◇ Reported Diamond
Occurrence

However, the relationship between the gold and gemstone-bearing deposits and the sub-basaltic sediments is uncertain.

Reports of diamonds discovered on the Toombullup goldfield appeared in the *Argus* and the *Mansfield Courier* in May 1896. According to the *Mansfield Courier* (9 May), a pea-sized crystal was found in a dish of wash taken from a tunnel driven on the Royal Extended claim, worked by Drummond and party. The crystal was sent to the Minister for Mines, who replied that a Melbourne diamond merchant had identified it as being similar to diamonds then being found in New South Wales. Several other possible diamonds had been found nearby. The Drummond claim was either at 'Lanky's Gully' or Coghill's Paddock, near Puzzle Gully, at the head of Dogwood Creek.

The region is popular with gem fossickers, particularly at Middle Creek and Webbs Creek. There have been sporadic reports of small diamonds being found but, until recently, these have remained uncorroborated. However, the discovery by Scott and Paul Bennett of a small (0.05 carat) tetrahexahedral diamond crystal in gravels at Webbs Creek in 2008 has lent authenticity to the 1896 reports. Pyrope garnet, considered an indicator mineral for diamonds, occurs in the Middle Creek gravels.

2.23 (*top*) Diamond from Webbs Creek, Toombullup district, discovered in 2008 (0.05 carats, 2.5 mm across).

2.24 Diamonds said to be from Diamond Creek (larger is 0.3 carats, 4.5 mm across).

KONGBOOL

The discovery of diamonds on the Congbool (later Kongbool) goldfields, in far western Victoria, was reported in the *Argus* on 27 April, 1895. The report stated that unnamed experts in Melbourne had confirmed small pale-yellow stones as diamonds. Sapphires and rubies were also found. The assistant Government Geologist, James Stirling, later reported in 1898 that the location was on the western side of Mathers Creek, about 2.5 miles (4 km) from the Kongbool Station homestead, on Allotments 56 and 57 in the Parish of Kongbool. The site is about 7 km WSW of the present-day town of Balmoral. According to a Mr Robertson, a small stone was found in situ by one of a group of miners sinking shafts through Pliocene gold-bearing gravels which cap the metamorphic rocks in the region. The finder gave the stone, along with some quartz crystals, to Robertson, who showed it to Stirling. The diamond was described by Stirling as a perfectly formed, pale-yellow octahedron measuring 1.5 lines (about 3.2 mm) from apex to apex and with adamantine lustre. Other small fragments with curved edges, perhaps resembling diamonds, were found in the deposits.

The whereabouts of any of the Kongbool diamonds are unknown. However the descriptions of their features and the geological setting of the discovery site resemble the Carapooee occurrence (see below) and suggest the discovery was authentic. The workings are now on Grassgunya property and consist of a few shallow trenches cut in red-brown quartz pebble conglomerate. Several mining companies, including Ashton Mining, CRA Exploration, and Continental Resources, investigated the locality during the 1980s (Morand *et al.*, 2003), but there are no published reports of further diamond discoveries.

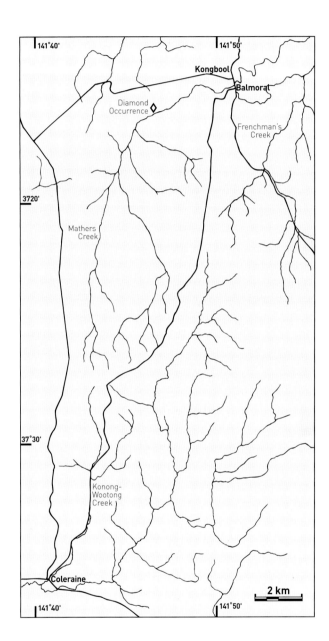

2.25 Map of Mathers Creek — Konong Wootong Creek district, western Victoria.

CARAPOOEE

Carapooee is a road junction about 12 km SSE of St Arnaud, an historic gold-mining town located 200 km northwest of Melbourne. St Arnaud prospered following the discovery in 1854-5 of rich gold in quartz reef systems extending for several kilometres to the northwest of the town. Some alluvial gold was obtained from shallow leads near the town during the 1860s. Gold was discovered in the Carapooee area late in 1854 and the goldfield was rushed several times between 1856 and 1858 (Flett, 1970). The field became known as Peters' Diggings, after the licensee of the local station, and consists of isolated workings in gravels which form cappings on low hills in the region.

The gravel deposits are believed to be remnants of an early Cenozoic river system and may be up to 8 m thick. The gold is concentrated near the base of the deposits, in a 0.5 to 1 m thick conglomerate layer consisting of rounded white quartz pebbles. This sits directly on highly weathered, metamorphosed Cambrian sedimentary rocks forming the basement. Zircons and sapphires of various colours (see Chapter 3), as well as abraded crystals of spinel, andalusite, and schorl, occur with the gold in heavy mineral concentrates (Birch *et al.*, 2007).

◊

2.26 Diamond found at Carapooee (0.5 carats, 4 mm across).

2.27 *(opposite)* Simplified geological map of the St Arnaud district.

Intermittent mining has occurred on various leases in the Carapooee–Kooreh district since 1989, firstly by Ando Minerals NL, which produced about 200 ounces (5.7 kg) of gold from the South Carapooee deposit before abandoning it in 1990 (Louthean, 1991). Kinex Mining Pty Ltd obtained approval to mine the Carapooee deposit late in 1990, but sold it in 1993 to Southern Border Gold Ltd. A small plant was established to process heavy-mineral concentrates, but recovery rates were uneconomic and the company went into liquidation in mid-1996. Late in 1993, a diamond was found on the vibrating table in the plant by Nigel Grigg, a student working at the mine. The diamond is pale yellow, measures 4 x 3.5 x 3 mm and weighs 0.48 carats. It is a slightly flattened tetrahexahedron with rounded outlines and a possible growth twin plane at one apex. While bright, the diamond appears abraded due to abundant percussion marks and curved fractures on its faces.

Precisely where in the Carapooee deposit the diamond originated is not known.

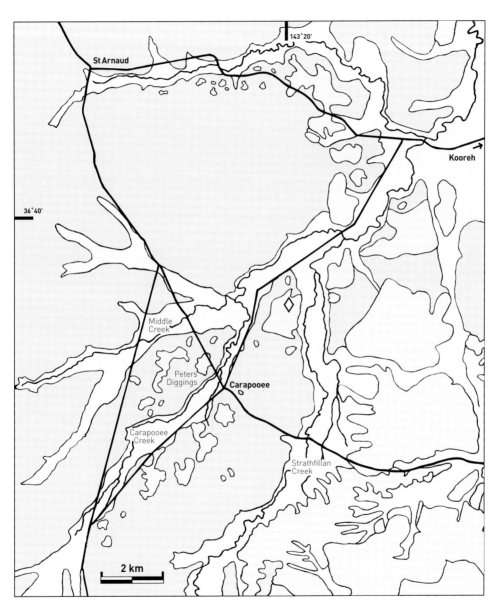

St Arnaud

143°20'

Kooreh

36°40'

Middle
Creek

Peters
Diggings

Carapooee

Carapooee
Creek

Strathfillan
Creek

2 km

Alluvium

Sand, gravel
(Tertiary)

Granite

Sandstone, shale
(Early Paleozoic)

◇ Diamond
Occurrence

DIAMOND CREEK

This stream is a small tributary of the Yarra River, about 25–30 km northeast of Melbourne. With other creeks it formed part of the Caledonia Diggings, where rich alluvial gold was 'officially' discovered in 1854 (Flett, 1970). The creek was named for the clarity of its water, allowing crystals to be seen on the bottom (Saxton, 1907; Edwards, 1979). Two small diamonds said to have been found in Diamond Creek were obtained in the 1970s by Mr Cyril Kovac, a Melbourne mineral dealer, and later donated to the Museum of Victoria. It is believed the stones were found when the creek was being worked for gold, possibly about 1930 during the Great Depression. One of the diamonds is a pale-yellow, flattened tetrahexahedral crystal, measuring 4.5 x 3 x 2 mm and weighing 0.32 carats. It shows one growth twin plane, strong frosting and prominent euhedral pits up to 0.5 mm across. These features are similar to those shown by diamonds from Bingara, New South Wales. There are no reports of other gem minerals, such as sapphires or zircons, from Diamond Creek (another Diamond Creek, east of Gembrook, drains into the Latrobe River, but there is no history of mining on this creek).

VAUGHAN

Vaughan, a small settlement on the Loddon River about 20 km north of Daylesford, was listed as a diamond locality by Perry and Perry (1997), but without any further information. The record may be based on reports of a diamond having been found during the 1970s by the lease-holder of the Vaughan Springs picnic reserve, which has been a popular sapphire locality with gemstone fossickers. While the occurrence cannot be completely discounted, it has been impossible to obtain any more details or to trace the supposed diamond.

OTHER LOCALITIES

The early reports of diamonds found at Yackandandah, Yandoit, Huntly near Bendigo, and Abbotsford in Melbourne remain doubtful, although the Yackandandah and Abbotsford stones were supposedly confirmed by experts at the time.

Late last century there were rumours of diamonds being found at Lake Bullenmerri, near Camperdown. This is one of the largest and most complex maar volcanoes in southwestern Victoria. Because these volcanoes erupted very explosively, they have been considered as possible locations for diamonds. Extensive sampling of the tuffs and beach gravels at Lake Bullenmerri was undertaken by geologists working for CRA Exploration in the late 1970s. Examination of the samples suggested that one contained diamond fragments. Re-sampling at Lake Bullenmerri failed to reveal further diamonds and it was subsequently shown that the original sample had been accidentally contaminated.

2.28 Simplified diagram for formation of 'subduction-related' diamonds in eastern Australia.

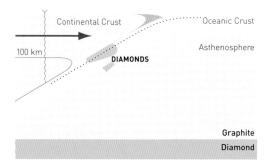

Origin of Victorian diamonds

The Victorian diamonds clearly form part of the eastern Australian diamond province, but with much lower concentrations compared to the occurrences at Copeton, Bingara and Oberon in New South Wales. Despite extensive exploration, primary sources for the diamonds remain undiscovered. Most prospectivity reviews (e.g. Hunt, 1979; Tan, 1982) and exploration programs for Victorian diamonds have assumed that one or more kimberlite pipes have released diamonds into the alluvial deposits. The highest concentration of diamonds in the Beechworth district suggests that this is the most likely location for a primary source. However, widespread multiple sources would be needed to account for the scattered occurrences of diamonds across Victoria. The search for indicator minerals, particularly in the Beechworth–Eldorado region, has not yielded conclusive results. Based on heavy mineral assemblages in the known Victorian diamond localities, the best 'indicator minerals' are blue to green sapphires. This association has recently been recognised for occurrences in New South Wales (Sutherland, 1996).

A model for the origin of the diamonds has to take account of the continental crust of eastern Australia being much thinner than crust normally associated with kimberlites. In other words, might it be possible for diamonds to crystallise at shallower depths than previously believed? In a new model, Barron et al. (1996) suggested that the diamonds may have formed in a slab of oceanic crust undergoing subduction. For diamonds to crystallise, temperatures in parts of the descending slab must remain cooler than in the surrounding rocks. This temperature difference will depend on the thickness of the slab and how fast it descends. As well, abundant

organic carbon must be present in sediments dragged down with the slab. Given all these conditions, diamonds could crystallise at depths of between about 80 and 100 km, about half the formation-depths associated with diamonds in kimberlites. Under this model, the diamonds in eastern Australia must have formed at least 250 million years ago, as this is the youngest known period of subduction affecting the region. Diamonds preserved in the crust could have been picked up and carried to the surface by magmas at various times during the widespread Mesozoic to Cenozoic volcanic activity in the region. Subsequent erosion of diamond-bearing lavas or associated volcaniclastic rocks would release the stones into the alluvial systems. The diamonds could then be recycled from one drainage system or alluvial deposit to another and spread widely and thinly over time. Sapphires and zircons brought to the surface during the same volcanic activity would be expected to show a similar distribution to the diamonds.

While the association of diamonds with sapphires in Victoria is not at odds with this model, more evidence is required on the age of the diamonds.

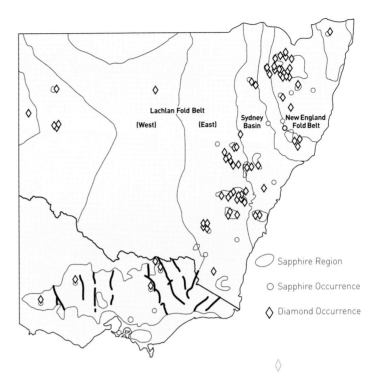

2.29 Distribution of diamond and sapphire fields in New South Wales and Victoria.

Diamonds in other Australian states

A great variety of diamonds has been found in alluvial deposits in all Australian states, but only in Western Australia's Kimberley region and at Merlin in the Northern Territory have they been found in large numbers in their host volcanic rocks, lamproite and kimberlite. Alluvial diamonds were abundant enough to be mined intermittently in the Bingara–Copeton district of New England, New South Wales, with smaller occurrences along the Macquarie River and in the Central Highlands. Diamonds are rare near Anakie, Gilberton and Stanthorpe in Queensland; in the Mount Donaldson district in Tasmania; and in the Echunga goldfield in South Australia.

2.31 *(top left)* Diamond crystal (13.2 carats) from the Argyle AK1 pipe, Western Australia.

2.30 *(bottom left)* Range of coloured diamonds (largest is 3 mm) from Argyle (Rio Tinto Collection).

2.32 *(below)* Heart-shaped pink diamond (1.2 carats) from Argyle set in a ring (Private collection).

ARGYLE

Exploration for diamonds began in the Kimberley region of Western Australia in the early 1970s. Geologists with Ashton Mining Ltd found small diamonds in creeks in the region and traced them upstream to find the sources, which turned out to be volcanic pipes consisting of lamproite. A pipe at Ellendale was not considered economic to mine at the time, but one found in 1979 on Smoke Creek, near Lake Argyle, was far more promising and was named the AK1 pipe. After several years of alluvial mining, the main Argyle Diamond mine on the AK1 pipe was opened in December 1985 and is now the world's largest diamond source. The operating company, Argyle Diamond Mine, is a subsidiary of Rio Tinto. Diamonds are liberated from crushed lamproite ore mined from an open cut and recovered on site, before being shipped to Belgium for grading and sale. Diamond production has averaged about 35 million carats (7000 kg) annually. Original reserves at Argyle were estimated as 5% (by weight) high-quality gemstones, 40% cheap gems and the remaining 55% industrial diamonds. Currently, the entire diamond production is cut and polished for the gem trade. Approximately, 72% are brownish diamonds marketed as 'champagne' and 'cognac' with 27% colourless to light yellow. Pink diamonds are extremely rare, accounting for less than 0.01% of production at Argyle. Indian cutters process most of the stones into gems. Pink diamonds are extremely rare and fetch very high prices (as much as $1,000,000 per carat) at annual auctions. There is also a collectors' market for Argyle diamond crystals.

Ore grades and reserves at Argyle are unusually high, making it economically feasible to continue mining. The company is planning to extend mining using underground methods.

Production commenced on the Ellendale pipe in 2003 and the mine now yields world-famous fancy yellow diamonds.

2.33 Diamonds from Argyle showing various colours, forms and surface features (largest is 4 mm) (Rio Tinto Collection).

COPETON

Alluvial diamonds occur in many places in eastern New South Wales but historically mining has been concentrated in the Bingara–Copeton district, near Inverell in New England, since the middle of the 1870s. The diamonds are mostly found in cemented gravels beneath basalt flows, and it has proved difficult to find a source. There are clear differences between the New South Wales (and other eastern Australian) diamonds and those from Argyle, reflecting different origins. Copeton diamonds are mainly pale-yellow dodecahedral crystals with a rounded appearance and weighing less than 0.5 carats. Twinning has made them harder than usual to cut.

2.34 The Argyle Pink Jubilee diamond was found in August 2011. At 12.8 carats uncut, it was the largest pink diamond ever found in Australia, but was reduced to 8 carats when cutting was halted. (Photo courtesy of Rio Tinto).

2.35 *(opposite)* Typical diamonds (up to 6 mm) from Copeton, New South Wales.

Sapphires, Rubies & Zircons

"Whatever may be said of the perfection in size and colour of particular stones, it nevertheless does appear that in no known country has there been brought together so great a variety of sapphires."

REV. JOHN BLEASDALE, 1864

◇◇◇◇◇◇◇◇◇◇◇◇◇◇◇◇◇◇

Sapphires — the transparent, coloured gem varieties of corundum — are widespread, even locally abundant, in Victoria. They occur as waterworn crystal fragments in many present-day streams, and in older gravels deposited by river systems earlier in the Cenozoic period. Blue is the dominant colour, but green, purple, yellow and brown varieties have been found. Pink stones, which may be called rubies, occur at some localities, but are much rarer and smaller than sapphires. Unlike occurrences in the New England district of New South Wales and the Anakie district in central Queensland, Victorian sapphires and rubies are too small and scattered to have supported commercial production. Not so widespread in Victoria are rounded pebbles of opaque corundum, with colours including white, grey, black, and pink. These may occur with sapphires in some deposits, but are believed to have a different origin. They are described in more detail in Chapter 9.

Crystals of zircon very often occur with the sapphires in the Victorian deposits, although there are localities where zircons occur without sapphires. Zircons show a wide range of attractive colours, from pale yellow, pink and lilac to orange, brown and deep blood-red, and more rarely blue and green. They also come in a diverse array of crystal shapes and rounded fragments.

Rounded crystals of black spinel are frequently associated with Victorian sapphires and zircons. Despite several early reports (for example by George Stephen in 1854), no 'ruby spinel' crystals have been found during recent sampling or in museum collections. It is possible the early descriptions were of red bipyramidal zircon crystals.

Most of the sapphires and zircons were originally enclosed in volcanic rocks, which were erupted throughout the Cenozoic period. These were either lavas, of which there are two main types — basalt and trachyte — or volcaniclastic rocks, which are fragmental rocks formed by early explosive eruptions. Weathering of the volcanic rocks releases the sapphires and zircons into soils and they become concentrated in river and stream gravels. Over time, the crystals can be carried very long distances because their hardness makes them resistant to abrasion. In some alluvial deposits, several different types of corundum and zircon occur together, suggesting crystals with different origins have been mixed and recycled.

This chapter deals with sapphires, rubies and zircons in groups of localities which have some geological features in common over a particular region. Because of their microscopic size, zircons derived from granites and old sedimentary rocks are not discussed. Nor are minor occurrences of corundum and zircon in beach sands along the Victorian coast (see Baker, 1962; Bell, 1953).

◊

3.1 *(opposite)* Panning for gemstones in Donnellys Creek, Gippsland.

Early history of discovery

SAPPHIRES

Victorian sapphires excited attention at about the same time as the diamonds, as both gem minerals were encountered during the earliest mining of gold-bearing gravels. Perhaps the first record was the waterworn, blue and white Ballarat sapphire shown by George Stephen during his address to the Geological Society of London in 1854. The most unusual discovery was made by the Melbourne jeweller, George Crisp, who told Rev. John Bleasdale that he had found a dark-blue sapphire crystal in the craw of a wild duck. George Ulrich's description of blue, green and magenta sapphires and red zircons from Guildford in 1860 was probably typical for the time, when diggers sent samples to the experts for their opinions. In the early 1860s, the sapphires were seen as the great hope of an established Victorian gem mineral industry. For several years, Bleasdale was the chief promoter of this cause, beginning with his talk to the Royal Society in November 1863, at which he exhibited blue, green and star sapphires from the Ovens goldfields. The 1865 exhibition of colonial gems, sponsored by the Royal Society, included sapphires of many colours from six localities. The Ovens goldfields yielded the greatest variety, from all shades of blue to green, yellow and brown, and including some star sapphires. Other localities featured were Dandenong Creek, Fryers Creek, the Daylesford region (Blue Mountain and Jim Crow Creek) and Gippsland. From 1865, sapphire and zircon discoveries, like those of diamond, were considered important enough to be published annually in *Mineral Statistics of Victoria*. While the Beechworth district continued to be the main source, new locality records, such as Tubba Rubba (Tubbarubba) Creek and the head of Woori Yaloak River (now

Woori Yallock Creek), were gradually added until the last report in 1882. Unfortunately many localities were rather vague, such as 'Upper Yarra', 'near Mansfield', 'Yarra Ranges' and 'Dandenong Ranges', reflecting the scarcity of geographical information available at the time.

Although blue sapphires were the most abundant, it was the rarer green crystals (or 'oriental emeralds') which created the greatest interest. Several large green stones of faceting grade were recorded from the Donnellys Creek diggings in northeast Gippsland during the mid-1860s, including one which produced a 5.75-carat gem when cut. The Jim Crow Creek diggings near Daylesford also produced green stones, including one fragment weighing 17.9 carats, which found its way into Stephen's collection. Sapphires were to the fore in the Victorian Intercolonial Exhibition of 1866–7; the Commissioners had a green stone from Donnellys Creek cut specially for the exhibition.

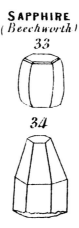

SAPPHIRE
(*Beechworth*)
33

34

In their respective essays for the exhibition, Bleasdale and Ulrich listed the main varieties of sapphire and their locations, with Ulrich providing the first sketches of small sapphire crystals from Beechworth. As well as the established localities such as the Beechworth and Daylesford regions, new occurrences of fine blue sapphires from the deep leads at Talbot and Mount Greenock were recorded.

3.2 *(opposite)* Sketches of sapphire crystals from Beechworth (Ulrich, 1866).

3.3 *(below)* Rounded sapphires (up to 8 mm) from the Eldorado Lead, Woolshed Valley, near Beechworth.

RUBIES

Rubies are transparent pink to red varieties of corundum, with more than about 1 weight % Cr_2O_3 responsible for the colour. Throughout the 1860s there had been occasional reports of 'oriental rubies' being found. In 1869, Robert Smyth reported seeing a ruby from the Daylesford district as early as 1853. George Ulrich discovered one in the Mount Eliza diggings, which he later described as a fine magenta colour. In 1870, a pea-size ruby was reported from the same gravel near Studley Park in Melbourne which had yielded the supposed diamond in about 1863. Bleasdale announced the discovery of rubies from near Pakenham in 1868; these were later stated to have been found in William Wallace Creek during prospecting for tin (Ulrich, 1870; Bleasdale, 1876). Edward Dunn described rare small grains of pale 'oriental ruby' from the Beechworth area in 1871. According to the Mining Registrar for Traralgon, in the Gippsland mining region, rubies had been found in a tributary of Traralgon Creek in 1876. An opaque pinkish variety of corundum was common as rounded pebbles in the Eldorado lead near Beechworth. Stephen gave it the name 'barklyite', in recognition of Sir Henry Barkly, the Victorian Governor from 1856 to 1863.

Although there are few analyses of these pink stones, it is unlikely that many are true rubies. Those with a purplish-pink tinge may contain more iron than chromium and are best referred to as sapphires.

◊

3.4 Sketches of zircon crystals (Ulrich, 1866).

ZIRCONS

Zircons were amongst the earliest gem minerals widely reported from the gold drifts. George Stephen told the Geological Society of London that he had been shown zircons from the Ovens District in August 1853, only about a year after those goldfields had been opened. Despite zircons being smaller, less attractive in colour and of lower value as gems than sapphires, many miners may have thought the abundant red crystals were rubies. In the Mineral Statistics reports between 1865 and 1882, zircons were reported from many of the sapphire localities. A few examples of sands rich in small perfect zircon crystals were reported from streams in Gippsland. The main localities for zircon were listed by Ulrich in his essay for the Intercolonial Exhibition of 1866–7. Many were from unnamed streams and leads in the Daylesford–Trentham–Castlemaine area. Ulrich noted the general range in colour from blood-red (the variety 'hyacinth') to near-colourless ('jargon'), and also described crystals from Daylesford showing beautiful blue to green dichroism. As described and illustrated by Ulrich, crystals were usually simple bipyramids combined with one set of prism faces, although more complex forms occurred in the Beechworth field.

Z I R C O N   *(Blue Mountain)*

27

28

29

(Beechworth)

30

31

32

◊

3.5 The Woolshed Valley, looking south from Mount Pilot.

USES OF SAPPHIRES AND ZIRCONS

Some early Melbourne jewellers were quite willing to cut many of the best sapphires and zircons, although the full extent of the commercial trade in these colonial gems is unknown. The fine green sapphires from Donnellys Creek were especially prized. Cut sapphires and zircons were featured in both the Royal Society and Intercolonial exhibitions. Presumably incomplete lists published in Mineral Statistics show that Spink and Sons cut small numbers of sapphires and rubies from Beechworth and Gippsland between 1874 and 1876. The same firm cut 40 zircons in 1875. Just as for the cut diamonds, no surviving examples of these faceted early Victorian gems are known.

Localities

There are so many localities in which the sapphire–zircon suite has been recorded that it is not possible to describe them all. The main localities have been grouped into regions, based on geological features and stream systems.

NORTHEASTERN VICTORIA
Ovens District

Corundum and zircon are widespread in the creeks forming the Eldorado Lead system in this region, which was outlined in Chapter 2. These include streams rising in the Beechworth Granite as well as in the Ordovician sediments. To the north, the Chiltern–Rutherglen lead system appears to have been poorer in these minerals, although this may be a reflection of a lack of reporting.

As mentioned previously, the multi-coloured sapphires from the Ovens gold diggings set pulses racing amongst gemstone enthusiasts in Melbourne from the early 1860s. While Bleasdale and Stephen were their most prominent advocates, it was Edward Dunn who systematically collected and described the full range of sapphires and zircons from the various creeks in the region. He observed near-colourless to dark Prussian blue sapphires as small rounded grains in most creeks, with green ('oriental emerald') varieties equally as widespread but less abundant. Pale pink ('oriental ruby'), pale purple ('oriental amethyst') and colourless to pale yellow ('oriental topaz') were much rarer, limited to a few small grains. An unusual variety was 'adamantine spar', described by Ulrich (1867) as hair-brown in colour with a silky opalescent lustre on the main cleavage plane. Star sapphires large enough to be cut for ring stones also occurred. According to Dunn (1871), crystals were rare, although both Bleasdale (1867) and Ulrich (1867) described yellow, greyish-blue, colourless and pale-blue prisms.

Sapphires similar to the Beechworth varieties were found in the Lancashire Lead to the north (Dunn, 1871), while blue and white stones, and occasional rubies, were reported by Dunn (1909a) as occurring in the Chiltern Valley Lead. Dunn also collected sapphires from the Indigo Lead.

Most of these varieties can be seen in Dunn's samples of Beechworth-district 'gem gravels' in Museum Victoria's collections. Generally, the sapphires appear as dull greyish-blue stones which are well rounded and between about 2 and 10 mm across. The stones are transparent to translucent, but almost always cloudy or flawed. A few fragments show faint colour banding. The rarity of gem-quality stones in Dunn's samples may be due to these having been selectively removed (for example, to give to George Stephen). There is a wide range of colours, with royal blue predominating, but pale or watery blue, purplish-blue, greenish-blue, yellowish-green and yellow varieties are also present. Greyish and pinkish-brown translucent stones also occur. Most sampled streams in the region show this distribution of coloured sapphires. Other Beechworth sapphires in the Museum's collections show a range of transparent to translucent coloured varieties, including pale greyish-purple and blue, bluish-green and greenish-yellow stones. There are also a few pale to deep purplish-pink stones, some of which may be rubies.

Zircons from the Beechworth district are extremely rounded, often bean-shaped and devoid of any trace of crystal faces. Most are transparent and range from dark blood-red through orange-red to pale-yellow, with some parti-coloured grains. Near-colourless to pale-pink stones are prominent. Yellow and orange stones may be difficult to distinguish from rounded cassiterite grains. The zircon grains are mostly between 2 and 6 mm long, but some may reach 9 mm. As with the sapphires, the proportions of the different colours show little variation across the region.

Accompanying the sapphires and zircons are grains of cassiterite (in colours ranging from black to dark brown, red, orange and yellow), topaz, garnet (of several varieties) and tourmaline.

Source of the sapphires and zircons
The distribution and extreme rounding of the sapphires and zircons led Dunn to support George Ulrich's suggestion that

◊

3.6 *(opposite)* Reedy Creek, near Wooragee (1996).

3.7 *(top)* Rounded zircons (up to 8 mm) from the Eldorado Lead, Woolshed Valley.

3.8 *(below)* Faceted sapphire (1 cm high, 7 carats) from Beechworth district.

they were probably derived from volcanic rocks. However there are no remaining basalt lavas in the watersheds of the Eldorado Lead and it is possible that the sapphires and zircons are eroding out of the older Cenozoic deposits in the Woolshed Valley.

Fission-track dates on the reddish zircons showed at least two populations, with ages of 5 and 19 million years. These dates do not match any known volcanic activity in eastern Victoria. The colourless to pale-pink zircons, which have low uranium contents (34 ppm), gave a much older date of 216 million years, in the Triassic period. This coincides with the age of a small province of syenite and trachyte near Benambra, about 120 km southeast of

◊ **3.9** *(below)* Sapphires (up to 1.2 cm) from Middle Creek, Toombullup district.

◊ **3.10** Sapphires (up to 8 mm) showing resorption, from Stringybark Creek, Toombullup district.

Beechworth. While these or related rocks of the same age may have provided the pale zircons, the evidence indicates a complex mix of sources for the zircon–sapphire suite.

Koetong Creek
Rounded pale-pink and red-brown zircons, with rare corundum, cassiterite and topaz, have been collected from Koetong Creek, in the far northeast of the State. This and other creeks which drain through the Koetong Granite contain alluvial tin deposits, worked intermittently between the 1870s and the 1960s (Cochrane, 1971). Pyrope garnets, with pale-yellow zircons (with some crystals of faceting grade), cassiterite and ilmenite were described from the Koetong area by Towsey (1979).

EAST–CENTRAL VICTORIA
Toombullup

At the time of the initial diamond discovery at this remote goldfield northeast of Mansfield, there were no reports of sapphires and zircons. However, as was noted by Kitson (1896), the miners were too busily engaged in getting the gold and utterly neglected the gems. Gold was obtained from shafts and tunnels dug into Cenozoic gravels both north and south of Dogwood Creek, a tributary of Ryans Creek, but sapphires, zircons and topaz were found only in the northern deposits, at Puzzle Gully (Kitson, 1896). Nowadays this area is covered by dense undergrowth and pine plantations, making access to the former diggings difficult.

Between 8 and 10 km to the east are the old workings on Webbs Creek and Middle Creek. The position of the former creek is uncertain, although it is likely to be a small north-trending tributary of Middle Creek. Kitson (1896) described the gold-bearing 'lead' at Webbs Creek as a stiff clay containing granitic and quartz–tourmaline pebbles, and a variety of gemstones. These included blue and green sapphires, red and colourless zircons, and topaz, as well as schorl, black spinel, 'rock crystal' and ilmenite. Many of the blue sapphires and red zircons weighed up to

3.11 Zircons (up to 12 mm) from Toombullup district.

◊

3.12 *(opposite)* Sapphires and zircons (red) from the Toombullup district (largest stone is 12 mm across).

3.13 *(top)* Faceted sapphires (0.67 and 0.64 carats) from the Toombullup district.

3.14 Faceted zircon crystal (7 mm long, 1.1 carats) from the Toombullup district.

2.5 carats and were flawless stones with good colour. The green sapphires were larger, up to 5 carats, but were extensively flawed. The workings on Middle Creek, about 1 km further east, were said by Kitson to be similar to those at Webbs Creek.

Despite its isolation, Middle Creek is a popular destination for gemstone fossickers. The main locality is about 800 m upstream from the Madhouse Road bridge, where there are remains of shallow shafts and sluicing pits. The gems can be obtained by panning samples from a thin layer of pebbly wash exposed in shallow excavations near the creek, as well as from the gravel in the creek bed. The Webbs Creek locality is not as well known to fossickers, but the nature of the gem minerals and their occurrence fit Kitson's description. Other creeks in the area, such as Stringybark Creek, Hollands Creek and Hoorihan's Creek (Ferguson, 1897), also contain sapphires and zircons. However, collecting is discouraged in those streams, which form the headwaters of Ryans Creek (which also contains these gemstones) because they are the main water supply for the town of Benalla.

Zircons are the most abundant of the gem minerals. They are mainly red and brownish-red, but greyish-brown, pale-brown, orange and pale-pink, transparent to translucent stones also occur. The fragments are well-worn and rounded, although crystal faces can often be made out. The largest stones reach 1 cm across. In most respects, these zircons are very similar to those from the Beechworth deposits.

The sapphires range from hexagonal cleavage fragments, barrel-shaped crystal sections and tapering points to irregular grains, ranging between about 3 and 12 mm across. While many grains are smooth and waterworn, there is a high proportion of rounded fragments showing etched-looking, dimpled surfaces which are probably due to magmatic resorption. Preservation of this feature suggests these crystals have not been carried far from their volcanic host rocks. The main colours of the sapphires are watery shades of royal blue, bluish-grey and greyish-green, with less common lilac-blue, bluish-green, yellowish-green and nearly colourless varieties. Amethyst-purple stones and pink rubies are extremely rare. Many of the sapphires are transparent, but most are translucent and flawed. Patchy colour-zoning is seen in some of the blue stones, and some nearly opaque dark-blue fragments show greyish stars. Several distinct varieties occur, including pale-greenish flawed fragments up to 1 cm across and rare examples of 'hair-brown' chatoyant stones.

The zircons and sapphires are accompanied by several different garnet types, including dark-red almandine and, notably, highly polished, transparent plum-coloured pyrope. Topaz, schorl, black spinel and fine-grained ilmenite may also occur within the gemstone concentrates. Cassiterite sand was noted with the gold in the original Toombullup goldfield (Stirling, 1898a).

Source of the sapphires and zircons
Overall, the Toombullup and Ovens District gem mineral assemblages are very similar. The Toombullup gemstones are less abraded, suggesting they are closer to their source rocks or have undergone less vigorous transport in streams. As at Beechworth, the gem minerals on the Toombullup field have mixed sources. For example, fission-track ages of 3.4, 20, 31 and 40 million years have been obtained on zircons from Middle Creek (Hollis *et al.*, 1986). The oldest of these dates coincides with the ages of the remnant basalt hill-cappings in the region (Wellman, 1974), but the sources of the younger zircons are unknown. The sapphires and oldest zircons may be eroding from sediments, possibly volcanic in origin, in former stream channels beneath the basalt cappings. Cenozoic gravels, such as those at Puzzle Gully, are another potential source. Gemstones released from these sources could also be incorporated into sands and gravels in present-day streams such as Middle Creek and Stringybark Creek.

Violet Town

Sapphires were found during road-works for the Hume Freeway near Violet Town, 140 km NNE of Melbourne, in the early 1980s. Excavations for a culvert about 3.5 km southwest of the town uncovered white clay at a depth of about 4 m. Blue-green stones up to 5 or 6 mm across were found in the clay; one stone was faceted for a ring. The deposit occurs in the course of a small unnamed stream rising to the east in hills composed of Late Devonian volcanic rocks. The source of the sapphires may have been Cenozoic lavas capping the hills but since removed by erosion.

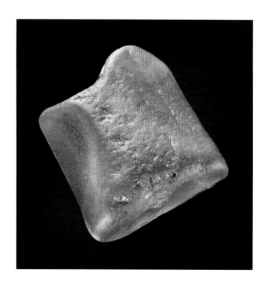

Donnellys Creek

Rich alluvial gold was discovered in Donnellys Creek, about 140 km east of Melbourne, in 1862 (Flett, 1970). It forms part of the remote Aberfeldy Goldfield, in a thickly forested, mountainous region of northern Gippsland. The creek is an eastern tributary of the Aberfeldy River and drains the southwest slopes of Mount Useful. Gold was recovered from wash along a 10-km stretch of the creek, then from nearby quartz reefs (Baragwanath, 1925b). Sapphires were reported from the creek in the mid-1860s, with green stones particularly notable. The largest, described by Rev. Bleasdale in 1866, was faceted by James Spink into a perfect 5.75-carat oval stone, 'well-proportioned, with a single cut on the bizil (sic) and four rows of facets on the collet side'. It was of a soft, even, chrysoberyl-green, with very good lustre and fire. A green Donnellys Creek sapphire was obtained by the Natural History Museum in London from Bleasdale in 1873.

The course of the creek is entirely in folded Devonian and Ordovician sedimentary rocks. A small patch of basalt (Older Volcanics) overlying Cenozoic conglomerates and gravels forms a capping on Mount Useful, and represents an earlier land surface. Terrace deposits containing quartz conglomerate and pieces of basalt occur along the course of Donnellys Creek (Baragwanath, 1925b). These were probably derived by erosion of the basalt and conglomerate on Mount Useful and are likely to have contained the sapphires.

Walhalla district

A 'hair-brown' sapphire from Stringers Creek, Walhalla, was exhibited by Rev. Bleasdale at the Philadelphia Exhibition in 1876. Rounded, brown to brownish-red zircon crystals up to 2 cm across have been found in the Thompson River near Walhalla. In one Museum Victoria sample, the zircons are accompanied by waterworn fragments of topaz and almandine.

◊

3.15 *(opposite)* Waterworn green sapphire (5 mm) from Donnellys Creek, obtained by the Natural History Museum, London, in 1873, from Rev. John Bleasdale (Collection of Natural History Museum, London).

3.16 Simplified geological map of Donnellys Creek area.

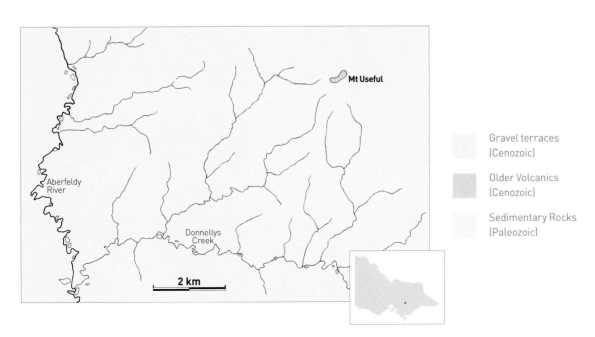

Gravel terraces (Cenozoic)

Older Volcanics (Cenozoic)

Sedimentary Rocks (Paleozoic)

Waterworn crystal fragments of zircon up to 7 mm, ranging from near-colourless to pale-pink and reddish-brown have been found with topaz, black spinel and gold in concentrates from the Tyers River 'between Goulds and Mt Baw Baw'. Gem clubs have reported sapphires from Jacobs Creek in the Erica district.

Dandenong Ranges

Several gem-bearing streams have their headwaters near the townships of Belgrave and Emerald, in the southern portion of the Dandenong Ranges, about 40 km east of Melbourne. The main streams draining the region are: Woori Yallock Creek, which flows north to join the Yarra River near Healesville; Monbulk Creek, which flows east to join Dandenong Creek; and Cardinia Creek, which flows south, but is now dammed by the Cardinia Reservoir.

The Dandenong Ranges consist of a thick sequence of Upper Devonian volcanic rocks on a basement of folded Paleozoic sedimentary rocks. A large granite mass (the Lysterfield Granodiorite) has intruded the volcanic rocks south of Belgrave. Remnant patches of basalt belonging to the Cenozoic Older Volcanics are preserved east and north of Emerald and in places on the granitic hills to the south. The gemfield closely coincides with the original Emerald goldfield, near the junction of Menzies and Woori Yallock creeks (Flett, 1970). Alluvial gold was discovered there in 1851, but the main rush took place in 1859. It appears several small tributaries of the Woori Yallock Creek (or as it was then known, amongst several names, as 'Woori Yaloak' and 'Moondie Yallock') were the source of the gold.

Unfortunately, there are no detailed published reports on the sapphires and zircons found in the area. The earliest official record is that of

Robert Smyth in 1872. He reported fire opals, sapphires, garnets and other precious stones from the head of the 'Woori Yaloak River' (nowadays probably Sassafras Creek), fragments of sapphires from the bed of a tributary of the Little Yarra River, and fragments and crystals of sapphire and corundum from the Dandenong Ranges. In view of the vagueness and ambiguity of the early place names, these records may all refer to the Emerald goldfield region. And while it is tempting to conclude that the goldfield itself was named because the gemstone 'emerald' was discovered in the creeks, it appears the name came from an early prospector, Jack Emerald, who was murdered there (Blake, 1977).

Earlier, in 1865, the Geological Survey had displayed 'an abundance of sapphires' from Dandenong Creek at the Royal Society's exhibition (Bleasdale, 1866a). Where on the creek's long course the gems were found was not revealed. Above its eastern tributary, Monbulk Creek, the headwaters include Selby and Hardys creeks, in both of which blue and green sapphires, as well as zircons, spinel, ilmenite, garnet and agate, have been found. Today, most of the former Emerald goldfield is either privately owned or within the Dandenong Ranges National Park, so access to creeks is difficult.

There are few samples of sapphires from the area to examine. However, some from the Belgrave–Emerald area are up to 1 cm across, are irregular in outline and show a colour range from pale watery-blue to green.

◊ **3.17** Simplified geological map of the Dandenong Ranges — Gembrook region.

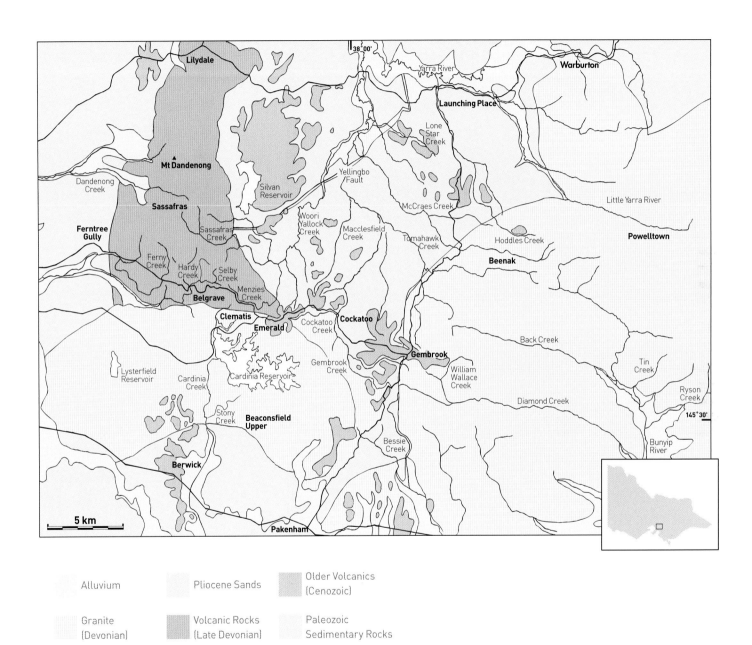

Alluvium

Pliocene Sands

Older Volcanics
(Cenozoic)

Granite
(Devonian)

Volcanic Rocks
(Late Devonian)

Paleozoic
Sedimentary Rocks

Cardinia Creek

Prior to the damming of Cardinia Creek in the late 1960s to form the Cardinia Reservoir, its northern headwaters were popular with gem fossickers (Smith, 1967a). The waters of the reservoir have now inundated the main collecting sites. According to Stone (1967), the drift of the creek, upstream and downstream of the former Wellington Road bridge, contained abundant chalcedony pebbles, accompanied by sapphires and zircons. These were usually concentrated in crevices between granite boulders in the creek bed. However, they could also be found in a buried layer of wash on nearby river flats.

Cardinia Creek sapphires were described by Smith (1967a) as generally of good colour and small, although stones up to 5.5 carats had been found. Sapphires made available by collectors consist of rough to waterworn chips and cleavage fragments, as well as irregular resorbed fragments, between 3 and 8 mm across. They show a wide range of mostly watery shades, from near-colourless, pale blue, bluish green, greyish blue, greyish green and greenish yellow to pale purple and 'lolly' pink. The few intensely coloured royal blue, green, yellow or pink stones that have been found are of faceting quality. An unusual brownish chatoyant variety also occurs, similar

◊ **3.18** Sapphires and rubies from Cardinia Creek (largest is 1 cm).

to stones found at Bass River and Beechworth–Eldorado. Zircons, as colourless, pale-pink and brownish-red waterworn bean-shaped grains up to 5 mm across, were found with the sapphires, along with amethyst crystal fragments, small agates and rounded schorl crystals.

Stony Creek is a short, west-flowing tributary of Cardinia Creek that rises near Beaconsfield Upper and flows entirely within the Lysterfield Granodiorite. Its gravels contain a suite of minerals that include zircon, sapphire, ilmenite, spinel, chalcedony and traces of gold. An 18-carat, pale-blue sapphire, possibly the largest ever found in Victoria, was reported from the creek in 2003.

Gembrook–Pakenham–Berwick

Miners spread out into the mountainous, forested country north and east of Gembrook during the 1860s. But without maps or major settlements, let alone roads, it was difficult to pinpoint discoveries. There is an early record of a gemstone mining lease being sought by the Gembrook Mining Company, on behalf of Messrs Page, Godfrey and LeSoeuf, all of Melbourne. They were applying to work for gemstones, by sluicing, on an unnamed creek, running northwest, about 4 miles west of William Wallace Creek and 15 miles northeast of Berwick. In their application they proposed to call the creek Gembrook, which coincides with the stream that still bears the name. Albert LeSoeuf and the Rev. Bleasdale were both credited by Saxon (1907) with naming Gembrook, but the town of that name was not established until 1874. Zircons and blue sapphires from 'country lying between the northwestern branch of the Bunyeep [Bunyip] River and Cardinia Creek' were donated to the Industrial and Technological Museum prior to 1869. Sapphires and rubies 'from the neighbourhood of Pakenham' and shown by

◊

3.19 *(top)* Faceted ruby (1.3 carats, 6 mm across) and green sapphire (3.5 carats, 7 mm across) from Cardinia Creek.

3.20 *(below)* Sapphires from the Pakenham district (largest is 9 mm).

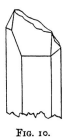

FIG. 10.

◊

3.21 *(top)* Sketch of 'ruby' crystal from William Wallace Creek (Ulrich 1870).

3.22 'Ruby' crystal (4 mm long), from William Wallace Creek.

Bleasdale at the Royal Society of Victoria in 1868, were said by Ulrich (1870) to have been found near the 'Berwick tin mine', on William Wallace Creek. In the account of his 1868 trip to William Wallace Creek with George Ulrich and Alfred Selwyn, Bleasdale wrote that their guide had found 'small bluish green sapphires, a few zircons and three small rubies, one of which was later cut'. In an important geological observation, the three men had confirmed that these gemstones were coming from the old decomposed basalt resting on the granite, which formed the head of the creek (Bleasdale, 1868a). No doubt drawing on these same observations, Ulrich (1870) noted that where the drift of the creek was full of nodules and pebbles of chalcedony, miners observed that coloured gemstones such as ruby and sapphire were usually present. There do not appear to be any official records of mining on William Wallace Creek, which may have been named for the 13th century Scottish hero, and the area remained unsurveyed until the early 1900s.

Ulrich (1870) described and sketched a 'ruby' crystal from William Wallace Creek donated to the National Museum by Bleasdale. Long thought to have been lost, it was eventually rediscovered amongst sapphires from another locality. Amethyst-purple and transparent, the crystal is a well-formed, but slightly abraded, hexagonal prism 4 mm long, with a broken pyramidal termination. A more recent description of the William Wallace Creek gemstones was given by Myatt (1972), who reported that small colourless, green and purple sapphires occurred with coloured chalcedony pebbles and yellow topaz fragments in the creek gravels.

Old catalogues and specimens in Museum Victoria show sapphires recorded from Gembrook, Pakenham and near Berwick. Only one creek name, Bessie Creek, about 4 km south of Gembrook, appears on a specimen label. According to a gem club report in 1976, small opaque cornflower-blue sapphires could be found 'in slippery pale grey clay in the stream bed'. This probably refers to the headwaters of Cockatoo Creek, just south of the Gembrook township, which has been the main locality visited by fossickers, to the present day. The sapphires from the Pakenham–Gembrook region, including the Cockatoo Creek stones, are mainly abraded, resorbed crystal sections between 2 and 6 mm in diameter, and tapering points up to 1.1 cm long. The colour is usually dark royal blue, with many sections showing greyish 'stars'. Dull grey to greenish stones are present. A few transparent fragments in shades of blue and green to near-colourless occur, some showing thin blue colour bands. In one Pakenham sample, the sapphires are accompanied by rounded, brownish-red to near-colourless zircons, black cassiterite crystals, green to brown xenotime crystals, and gold.

Source of the sapphires and zircons
Extensive patches of Cenozoic basalt in the region provide a potential source for the sapphires and zircons. Bleasdale (1868a) and Ulrich (1870) were the first to draw a link between the agates, sapphires and rubies in William Wallace Creek and small areas of basalt capping parts of the ranges. Layers of gravel and clay deposited in former stream beds occur in places beneath the basalt hill cappings (VandenBerg, 1971). These sediments below basalt were first observed by Murray (1884), then by Stirling (1899b) near the head of Cockatoo Creek in the Gembrook goldfield. The headwaters of William Wallace Creek and Bessie Creek also drain from this patch of high-level basalt on which Gembrook is situated. If the sapphires and zircons are being eroded out of sediments beneath the basalts, the gemstones may have

originally come either from nearby volcanic
vents or from distant basalt source rocks.
The age of the lavas and the sediments beneath
them is uncertain. The only dated lava in the
region is from Berwick and has an age of
22.3 million years (McKenzie *et al.*, 1984). Leaf
fossils beneath this basalt are of Eocene age
(between 37 and 48 million years old) and this
is possibly the age of the sub-basaltic sediments
in the Gembrook area (VandenBerg, 1971).

Upper Yarra region

This region covers some of the southern
tributaries of the Yarra River between the
present-day towns of Launching Place and
Warburton, and includes Hoddles Creek,
Britannia Creek, Scotchmens Creek, Little
Yarra River and Big Pats Creek. All of these
streams were worked for gold in the early
1860s (Flett, 1970), but there are few published
accounts of gemstones being found. Ulrich
(1867) referred to tin concentrates with

◊ **3.23** View over William Wallace Creek,
near Gembrook.

sapphire, schorl and gold in the Upper Yarra
Tin Company's claim. Further east, sapphires
occurred with black cassiterite sand in Tin Creek,
a tributary of the Bunyip River (Selwyn *et al.*,
1868). There are some small ruby crystals
labelled 'Upper Yarra' in the Museum Victoria
collection but their provenance is uncertain.

Other localities

Sapphires and zircons have been reported
from gravel in Russells Creek, near Hill
End, about 20 km northwest of Moe. Small
sapphires have also been reported from a
road-cutting a few kilometres south of Dookie.

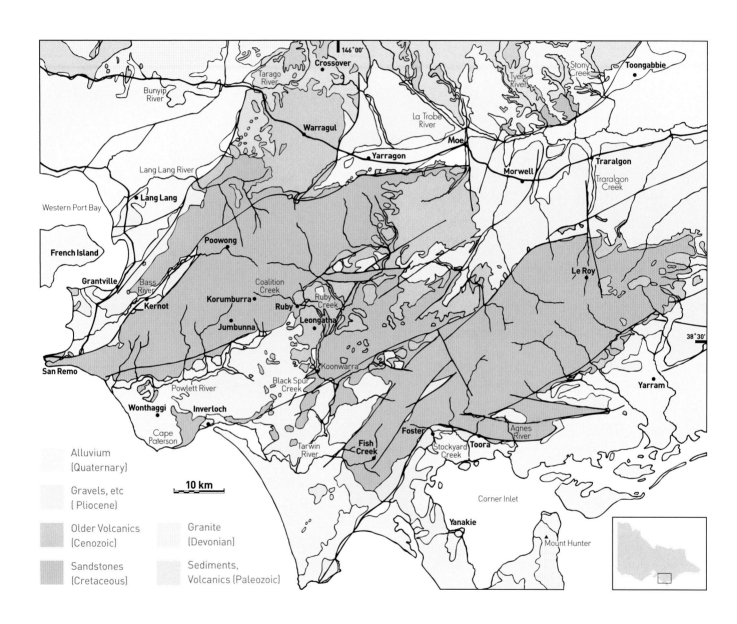

Alluvium
(Quaternary)

Gravels, etc
(Pliocene)

Older Volcanics
(Cenozoic)

Granite
(Devonian)

Sandstones
(Cretaceous)

Sediments,
Volcanics (Paleozoic)

10 km

146°00'

38°30'

Crossover

Tarago
River

Bunyip
River

Warragul

Lang Lang River

Yarragon

Moe

Morwell

Traralgon

Traralgon
Creek

La Trobe
River

Tyers
River

Stony
Creek

Toongabbie

Western Port Bay

Lang Lang

French Island

Grantville

Bass
River

Kernot

Poowong

Korumburra

Coalition
Creek

Ruby
Creek

Ruby

Jumbunna

Leongatha

Le Roy

San Remo

Powlett River

Cape
Paterson

Wonthaggi

Inverloch

Tarwin
River

Black Spur
Creek

Koonwarra

Fish
Creek

Foster

Stockyard
Creek

Toora

Agnes
River

Yarram

Corner Inlet

Yanakie

Mount Hunter

SOUTH GIPPSLAND

This region encompasses several scattered and poorly documented localities for sapphires and zircons, between about 80 and 160 km southeast of Melbourne. Gold was won from several alluvial systems in the 1860s and 1870s, but there were no particularly rich fields. The main goldfields were in the Traralgon district; near Crossover and at Stockyard Creek near Foster. The alluvial deposits were mostly worked out by the late 1870s and attention turned to coal deposits in the Mesozoic sedimentary rocks of the region.

The geological history of this region is complex. It forms the western edge of the Gippsland Basin, in which sediments have been deposited from the Late Cretaceous to the present day. These include the extensive brown coal deposits of the Latrobe Valley as well as the sediments which host the off-shore oil and gas deposits in Bass Strait. Movements on many faults and monoclinal folds have warped the sediments. Volcanic activity has occurred at intervals, with some lava flows buried in the sedimentary sequences. The South Gippsland Hills are up-faulted blocks of Lower Cretaceous sandstones and mudstones which form the basement of the region. Draped over these are extensive flows of basaltic lava belonging to the Older Volcanics. These are of two main age groupings, 49–57 and 19–24 million years old (McKenzie *et al.*, 1984, Wellman, 1974). Preserved beneath the basalts in places are deposits of gravels, clays, conglomerates and brown coal (known as the Childers Formation) of Late Oligocene age (Hocking *et al.*, 1988). During the Pliocene epoch, extensive sheets of gravels, sands and clays (known as the Haunted Hills Formation) were deposited over the region by streams flowing from the highlands to the north.

Bass River

The Bass River is one of the main streams draining the western part of the region. It rises in the highlands near Poowong and flows in a southwesterly direction, reaching Western Port Bay NNE of San Remo. Before entering a wide flood-plain, its course is through a fault-controlled valley in Lower Cretaceous sandstones which are draped on the east by Pliocene gravels.

In 1905, The Victorian Sapphire and Precious Stone Exploration Company applied for a lease covering 50 acres (20 ha) in allotments 141a and 156, Parish of Corinella, in order to mine

◊

3.24 (*opposite*) Simplified geological map of South Gippsland.

3.25 (*below*) Sapphires from Bass River (largest is 7 mm).

◊

3.26 (*above*) Faceted sapphires from Bass River (blue-green) and Foster (orange) (largest is 2.5 mm).

3.27 (*below*) Faceted sapphires from Bass River (largest is 7 mm).

sapphire and corundum (White, 1974). The gemstones had apparently been discovered during exploration for gold on the Campbell property near the township of Kernot. This is probably the occurrence referred to as being 'near Grantville' by Kitson (1902b). Some excavation work was carried out on the banks of the Bass River, but the company soon ran out of capital and closed its operations. No production records are available.

Edward Dunn appears to have collected extensively at the locality at some stage, and several of his gemstone concentrates are in Museum Victoria's collection. The sapphires are the most attractive of all the Victorian stones. Blues predominate, from deep inky blue to pale watery shades, many being sufficiently transparent and free of flaws to be of gem quality. The other main varieties represented are purplish red or 'plum', yellow to green including parti-coloured, pale green and dark greyish brown. Cleavage fragments of a 'hair-brown' chatoyant variety are prominent. These are larger (up to 1.2 cm across) and appear less abraded than the coloured varieties. Most of the gem varieties are waterworn but many stones show rounded, dimpled surfaces characteristic of resorption by magma. Other concentrates in the Museum's collection show this colour range in sapphires between 3 and 7 mm across, but with some bright pink rubies and pale-purple stones present as well.

◊

3.28 'Star' sapphires (blue) from Pakenham district and 'raspberry' sapphire (4 mm) from Bass River.

Larger stones are reported to have been found by members of local lapidary clubs. Pale-brown to red zircons and rounded grains of black spinel occur with the sapphires.

Gemstones can still be found in the Bass River, although the main collecting site near Kernot is on private property. The source of the gems has not been traced.

Foster

Gold was found in 1870 in Pliocene gravels capping several low hills (including Kaffirs Hill) near the township of Foster. The younger gravels in some of the local streams, in particular Stockyard Creek, were also found to be auriferous (Stirling, 1894). The gold was traced to nearby quartz reefs. Nicholas (1876) listed Stockyard Creek as a locality for sapphires and zircons, although he provided no descriptions. Samples in Museum Victoria's collection from 'the drift of a gully near Foster' and from 'Kaffirs Hill' contain a variety of sapphires. The most notable are transparent but flawed, pale-brown to pinkish-brown resorbed fragments up to about 7 mm across; this variety has not been seen at other localities. Other types present include dull blue, bluish-green, green and purplish, translucent to opaque, water-rounded stones between 2 and 10 mm across, and a few rough blue and 'hair-brown' cleavage fragments about 7 mm across. There are also several 2-mm rubies. Zircons are mainly small, near-colourless to orange and brownish-red, rounded crystals. Black spinel and rare topaz and cassiterite occur in the Kaffirs Hill sample.

Toora

About 15 km east of Foster is the Toora tinfield, where alluvial cassiterite was discovered in the 1870s. The deposits occur in Cenozoic gravels, sands and clays and were worked between about 1888 and 1914 (Cochrane, 1971). The Agnes River and Tin Mine Creek, an eastern tributary of the Franklin River, flow through the deposits. Smyth (1873) listed blue and white sapphires, corundum, zircon and topaz from the Franklin River, which drains the western portion of the tinfield. Murray (1890), Kitson (1901, 1902a) and Spencer-Jones (1955) described other heavy minerals including tourmaline, spinel, ilmenite, garnet and monazite in the gravels.

Lang Lang River

Bluish-green sapphires, as fragments up to 10 mm across, have been found in the Lang Lang River, near Lang Lang, with some stones being of faceting quality. The locality is well known to local gem fossickers.

Traralgon Creek

The Mining Registrar for the Traralgon subdistrict reported in December 1876 that rubies were being found in a tributary of the Traralgon Creek (Couchman, 1877). No further

3.29 Sapphires from Foster (largest is 6 mm).

details or specimens are known, but the locality may be near the township of Le Roy.

Yanakie

On the western coast of Corner Inlet, Yanakie is a small settlement just north of an outlier of the Wilsons Promontory granite. Murray (1876) noted that the granite contained small garnets which could be washed from detritus containing zircon, green and blue sapphire, topaz and 'almandine ruby'. These had been examined by George Ulrich. The gemstones were probably derived from a patch of Pliocene gravel overlying the granite.

Yarragon

This locality for sapphires was listed by Atkinson (1897) and Nicholas (1876) without further detail. A concentrate in the Museum Victoria collection contains mainly dull to navy blue sapphires, with a few greenish-blue and yellowish-green stones. They are transparent to translucent and up to 1 cm across. The sample also contains several bright pink rubies, between 2 and 3 mm across, and some pale zircons.

Toongabbie

Zircons from alluvial deposits in streams (such as Stony Creek) near Toongabbie (Stirling, 1899c) are a mixture of colourless to pale pink, orange and reddish-brown, transparent, rounded grains, generally between 2 and 4 mm across. They have a brilliant lustre and closely resemble the zircons from Diamond Gully near Castlemaine. They occur with black spinel grains and rare sapphire.

Other localities

Sapphires from the Tarwin River, from near Crossover (Creek), and in wash-dirt on Upper Boggy Creek were donated to the Industrial and Technological Museum prior to 1869, but have since been misplaced. Atkinson (1897) and Nicholas (1876) listed Latrobe River and Bairnsdale, and Krause (1896) listed Leongatha as sapphire localities, but without detail. Kitson (1902b) reported that prospectors had found coal seams while searching for 'rubies' and sapphires in Coalition Creek. He suggested that Ruby Creek and the town of Ruby, near Leongatha, were named for red zircons found in the creek by early settlers and prospectors. Murphy (1988) came to a similar conclusion, but referred to garnets, which are much less likely than zircons to occur in the region. Kitson (1905) listed Koorooman (the parish covering this area) as a source of sapphires. Gem-club reports record sapphires from Black Spur Creek, near Koonwarra, and from an unspecified locality on the Powlett River. There are also reports of sapphires being found in Smile of Fortune Creek and other streams near Moondarra.

Source of the sapphires and zircons

Volcanic rocks are likely to have been the primary source for the sapphires and zircons of southwest Gippsland. However, there is no obvious relationship between the gemstone locations and the present-day distribution of the lavas belonging to the Older Volcanics in the region. This may be due to the removal of extensive areas of basalt by erosion. At all the known localities, sapphires and zircons have been found in streams flowing through or close to gravel deposits of the Pliocene Haunted Hills Formation or its equivalents. This relationship between gemstone-bearing creeks and the 'Cainozoic pebbly drifts' through which they flowed was observed by Kitson (1902b). The gem minerals are likely to be concentrated in former stream channels within the gravel sheets, having originally been weathered out of the Older Volcanic lavas.

MORNINGTON PENINSULA AND WESTERN PORT

The Mornington Peninsula is a faulted block of Paleozoic sediments and granite separating Port Phillip and Western Port bays. Two main localities for sapphires and zircons occur on the Peninsula.

Tubbarubba Creek

This is a small stream on the Peninsula, about 50 km south of Melbourne. The creek drains NW for about 5 km, and is on private land. Alluvial gold was discovered in Tubbarubba Creek and nearby Bulldog Creek late in 1862, with several rich patches exploited. Quartz reefs intersecting possible indicator bands of dark Ordovician slate were later discovered and worked intermittently for gold. Sapphires and zircons were found with the gold in the alluvial deposits. George Ulrich collected a small 'oriental ruby' during a visit to the 'Mt Eliza diggings', as he referred to the field,

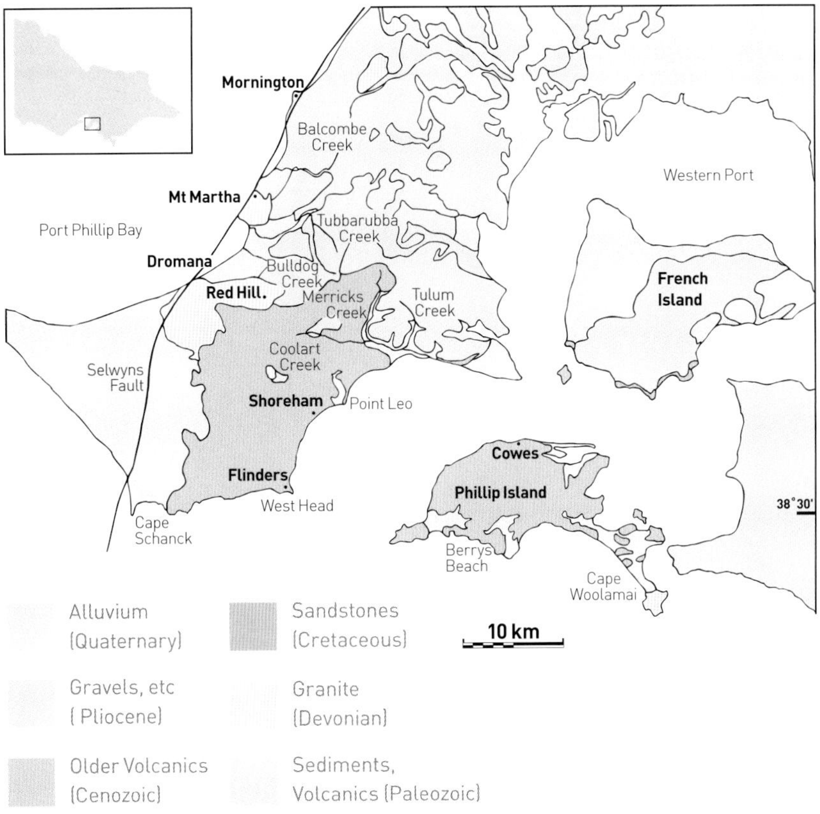
Mornington
Balcombe Creek
Mt Martha
Port Phillip Bay
Tubbarubba Creek
Dromana
Bulldog Creek
Red Hill.
Merricks Creek
Tulum Creek
Western Port
French Island
Selwyns Fault
Coolart Creek
Shoreham
Point Leo
Cowes
Flinders
Phillip Island
West Head
38°30'
Cape Schanck
Berrys Beach
Cape Woolamai
10 km
Alluvium (Quaternary)
Sandstones (Cretaceous)
Gravels, etc (Pliocene)
Granite (Devonian)
Older Volcanics (Cenozoic)
Sediments, Volcanics (Paleozoic)

◊

3.30 *(above)* Simplified geological map of the Mornington Peninsula.

3.31 Faceted sapphires from Tubbarubba Creek (largest is 7 mm).

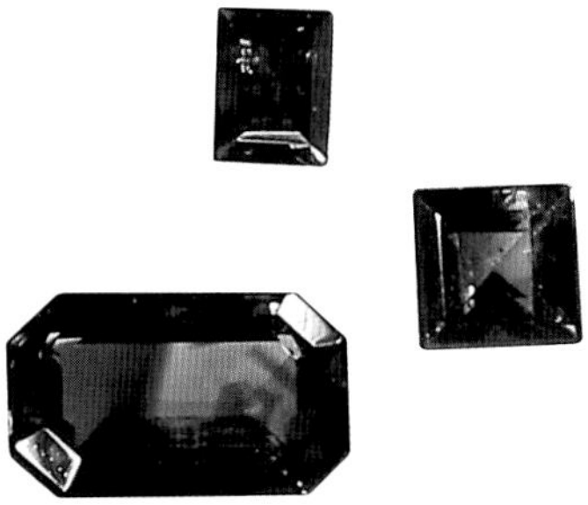

Tubbarubba Creek

The first gold claim I ever had
was at Tubbarubba Creek;
Once worked by early miners
in those harder days of old,

I had studied all the records and
heard old miners speak,
And hoped to match some earlier
find, in tales so often told.

The mining there was simple,
just a dish or rocking cradle;
Water there was far too scarce
for riffle boards and sluice;
I sometimes had to dig a hole,
and scoop water with a ladle,
Then seasons with a lack of rain
gave rest a good excuse.

Each dish I washed had residues
of colours blue and crimson,
And there, though all too sparse,
some minute specks of gold;
Those stones, alas! Too small
to cut, of sapphire and the zircon,
So were put in little bottles, to be
kept and sometimes sold.

Many years have passed away
since those early mining days;
I've tried in many places for
those minerals that we seek;
Memories of those days long
past, now gleam like sunlit rays,
And I often dwell in fancy, back
at Tubbarubba Creek.

ARCH BROWN
(*The Australian Lapidary Magazine*, April 1973)

◊

3.32 Sapphires from Tubbarubba Creek
(largest is 1.1 cm).

in about 1864. Although gold mining ceased soon after, fossickers were still obtaining sapphires by sieving the creek gravel in the 1960s.

Sapphires from Tubbarubba Creek are generally transparent, bright royal blue stones between 1 and 6 mm across. However the complete colour range of transparent gems includes deep greenish blue to pale watery blue and green, yellowish green, purplish blue and pale lilac. The colour is usually quite uniform. Translucent to opaque stones may show colour banding, intergrowths or overgrowths of blue on grey, and may be dull greyish purple or brownish grey. The sapphires range from well-rounded crystal sections to irregular waterworn fragments up to 1 cm across. Possible rubies are indicated by small pale to bright pink grains up to 3 mm across. The accompanying zircons are reddish orange, rounded grains up to 3 mm across. Smaller sapphires and zircons showing similar colours to the Tubbarubba Creek stones also occur in Bulldog Creek.

Point Leo

Point Leo (Point Bobbanaring) is a low promontory on the southeast margin of the Peninsula, about 8 km south of the Tubbarubba Creek locality. The Point and low cliffs fronting the sandy beach to the northwest consist of weathered Cenozoic basalt. The beach has been a popular locality for gem fossicking, but this activity is now banned due to holes dug in the sand being left unfilled. Sapphires and zircons are found in a layer about 1 m below the beach. They may also occur concentrated in the matrix of pebble conglomerates at the high-tide level on the beach, and on the rock platform of the point itself (Pauley and Smith, 1967). Rounded grains of black spinel occur abundantly with the gem minerals. Point Leo sapphires and zircons are generally smaller versions of the Tubbarubba

Creek stones, although Point Leo zircons up to 1 cm across and sapphires to 5 or 6 carats are known.

Sapphires and zircons have also been found in the sands at Tulum Beach, about 5 km northeast of Point Leo.

Phillip Island

Sapphires and zircons have been reported in beach sands on the south coast of Phillip Island, in Western Port Bay. The gemstones are undescribed and the locality is uncertain but is likely to be Berrys Beach.

Source of the sapphires and zircons
Thick basalt flows of the Older Volcanics occur on the Mornington Peninsula to the south of the Tubbarubba Creek headwaters. Merricks Creek and Coolart Creek have cut down through the basalt and flow southeastwards to reach the coast north of Point Leo. These streams may have carried the sapphires and zircons to the beach deposits. This distribution suggests there may be a source associated with the basalt near the main divide on the Peninsula. Point Leo zircons have been dated at 47–48 million years old, within the range of ages (38–59 million years) obtained for the basalts (Sutherland, 1996). There are some isolated patches of post-basaltic 'pebble beds' of Eocene–Oligocene age, which may be the remains of alluvial deposits in former valleys cut down through the basalt flows. Keble (1950) suggested that some of the

alluvial gold may have been derived from these nearby pebble beds. They have not been examined to see if they contain sapphire and zircon.

Thick basalt flows of similar age to the Mornington Peninsula lavas cover much of Phillip Island. Sapphires and zircons from a source beneath the lavas could have been carried to south coast beach sands by short streams such as Native Dog Creek.

3.33 Sapphires and zircons from Point Leo (largest is 5 mm).

MACEDON RANGES REGION

Many volcanic hills dot the region between Woodend and Lancefield. These erupted a complex series of basalt and trachyte lava flows between 6 and 8 million years ago. Important trachyte volcanoes include Hanging Rock, Brocks Monument, Camels Hump, the Jim Jim, Mount Kerrie and McLarens Peak. The trachytes contain abundant zircon crystals, which can now be found concentrated in many of the small streams in the region. Ilmenite predominates in the concentrates, with corundum, usually as grey or pale-mauve fragments, being very rare. Pink garnet, derived from the Devonian dacite forming Mount Macedon, is locally abundant.

At Camels Hump, on the summit of Mount Macedon, zircons up to 2 mm long can be found in soil pockets preserved on the rocky outcrops. A small, unnamed creek crossed by Sheltons Road, east of Newham, contains abundant small pink zircon crystals which show exquisite forms. Abundant amongst these are complex twinned groups of up to 20 crystals forming clusters resembling snowflakes. Similar zircon suites can be found in Bolinda Creek draining the Mount Kerrie trachyte, and in creeks draining the western slopes of the Jim Jim.

WEST-CENTRAL HIGHLANDS

The West-Central Highlands region, centred on the town of Daylesford, about 85 km northwest of Melbourne, is perhaps the most prolific source for sapphires and zircons in the State. The gemstones were first reported during the alluvial gold rushes of the 1850s, with George Ulrich providing the earliest descriptions of sapphires and zircons from the 'Jim Crow' goldfield in 1860 (*The Herald*, 18 September). Few early samples have been preserved in museum collections but, fortunately, it is still possible to collect gemstones by panning in many of the region's streams.

3.34 Gold mining under lava, Jim Crow diggings, in the late 1850s. La Trobe Picture Collection, State Library of Victoria.

The bedrock of the region, north of the east–west trending Greendale and Coimadai faults, is of folded Ordovician slates and siltstones. The north–south trending Muckleford Fault, a large reverse fault, is close to the western edge of the region. Apart from a few patches of Permian glacial sediments, the youngest rocks are sands and gravels deposited in earlier Cenozoic stream valleys and the lava flows which filled many of the valleys, especially in the northern half of the region. Elongated patches of basalt, often forming plateaus, mark the valleys of former streams filled by the lava flows. The buried streams form the eastern and southeastern extremities of the extensive Loddon Deep Lead Group (Canavan, 1988; Hunter, 1909). This system of leads covers a total area of about 2500 km^2 and includes most of the richest alluvial gold deposits in the State. The present-day drainage pattern, including the Loddon River, which rises east of Daylesford, shows only limited relationship to the ancestral system. The previous divide was further to the south, so that former streams in the Daylesford–Trentham region had their headwaters in the Ballan area and flowed north to join the ancestral Loddon River, bringing gemstones with them.

Detailed sampling of the sands and gravels in the region's streams by Dr Julian Hollis has shown zircon occurring in over 150 localities, about half of which also contain sapphires. The gemstones fall into two distinct age groups. Older zircons and sapphires from the deep lead deposits are often found with a wonderful array of zircon crystals derived from the young lavas of the Newer Volcanics.

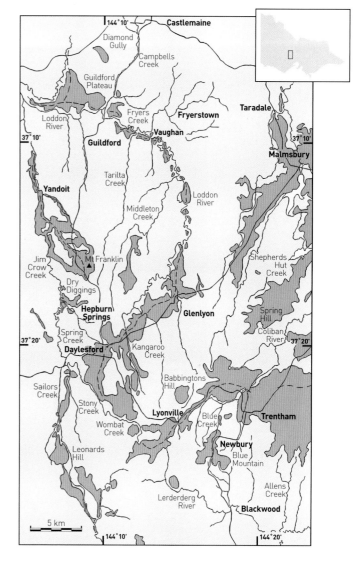

Basalt, trachyte (Newer Volcanics)

Sandstone, Slate, (Paleozoic)

3.35 Simplified geological map of the West-central Highlands region.

Daylesford district

The early gold diggings near Daylesford were called Jim Crow Creek or Ranges, and were along the downstream course of the main creek flowing northwards past the town. That part of the creek flowing from the south was named Sailors Creek. Wombat Creek, on which the earliest gold was found late in 1851 (Flett, 1970), joins Sailors Creek just south of the town. For most of its northern course, Jim Crow Creek is a marginal stream to lava flows from old volcanoes, centred on Mount Franklin, occupying what may be the stream's former valley. Cenozoic sands and gravels beneath the basalt form the Yandoit Lead, named for the town of Yandoit, from where a diamond was reported to have been found in about 1859. Gold was worked in Jim Crow Creek from about 1852 (Flett, 1970), and sapphires were being found soon after. Smyth (1869) reported that, as early as 1853, George Foord had discovered a waterworn, blue and white crystal and a ruby, of good colour but milky, in a parcel of stones from the Jim Crow diggings. Large green stones from Jim Crow Ranges, similar to those from Donnellys Creek in Gippsland, were exhibited by Rev. Bleasdale at the Royal Society (Bleasdale, 1867). George Ulrich described Daylesford sapphires as the State's largest, with cleavage fragments, hexagonal in outline and up to 1/2 inch (13 mm) across, not uncommon. The sapphires were a mottled or cloudy mixture of transparent deep blue, with patches of various shades of green and opaque grey or black. Ulrich (1867) also referred to fine uniform bluish-green stones, including one flawed stone nearly one cubic inch (16.4 cm³) in size belonging to George Stephen. This was probably the 17.9-carat fragment, 1.75 inches (4.4 cm) across the prism, which Stephen exhibited at the Royal Society in 1865 (Bleasdale, 1865). Sapphires from Daylesford in the Museum Victoria collections include stones

◊

3.36 Sapphires from the Daylesford district (largest is 7 mm).

from Edward Dunn's collection. Although they are not labelled as coming from Jim Crow Creek, the preponderance of green stones indicates this is their most likely source. They are rough, rounded fragments between 4 and 10 mm across, and generally appear dull greyish-green and brown. The dominant green stones are pale watery shades and quite flawed.

A few kilometres southwest of Mount Franklin were the Dry Diggings, which were opened in 1853 (Flett, 1970). The gold occurred in wash beneath a small patch of high-level basalt. The diggings were listed as a source of sapphires by Smyth (1871) but no details were given.

Sailors Creek, south of Daylesford, drains the young lava flow from the Leonards Hill volcano, as well as various deep lead remnants upstream to Sailors Falls. The creek is named for the party

of sailors who discovered its rich alluvial gold late in 1851. Zircon crystals from gravels in the creek were collected in 1855 by Henry Rosales, who misidentified them as vesuvianite because of their perfect tetragonal form; crystals of this perfection have not been found there since. The heavy minerals in the creek are dominated by young zircon crystals, ranging from pale-brown prisms which are often skeletal, to beautiful bright-red bipyramidal crystals. These may be stumpy or elongated to about 10 mm, depending on the length of the prism faces. A few larger, abraded, deep lead zircons also occur, with colours ranging from pale pink, yellow and mauve to scarlet. Sapphires are extremely rare.

Sapphires were more abundant in gravels in the Italian Hill Lead, which was opened in 1857 southeast of the town. It trends SE–NW towards the centre of Daylesford, and joins the basalt-covered lead system which runs for about 10 km in a SW–NE direction to Glenlyon, then turns northwards. Tailings dumps beside Patterson Street in Daylesford have come from tunnels under the basalt to the east. A gem-gravel sample from the Italian Hill Lead, collected by Dunn, shows a few pale to dark-blue and pale-green rounded sapphires up to about 5 mm across. They occur with dull greyish opaque corundum fragments up to about 9 mm across, abundant rounded zircons with colours from pale pink and yellow to brownish-orange and red, large rough fragments of black spinel, small cubes of pyrite and flakes of gold. Similar minerals have recently been collected from the Patterson Street dumps. Pale-blue sapphires to about 6 mm occur in gravels in Wombat Creek flowing from the southeast. Its course is close to the margin of the Italian Hill Lead, from where some of the gem minerals are derived.

◇

3.37 Zircon crystal (2.5mm) in basalt from Ridge Road quarry, near Daylesford.

Significant alluvial sites where gemstones can be collected near Daylesford include Spring Creek at the old Ballarat Road, and Jim Crow Creek at Tipperary Spring and downstream from Breakneck Gorge, Hepburn. These yield complex mixed samples similar to those from Sailors Creek. Scarlet to orange garnet and purple corundum occur with the sapphires and zircons.

An abandoned basalt quarry on Ridge Road, 3 km southwest of Daylesford, is an important site for zircon crystals. The basalt is a valley flow from the Leonards Hill volcano and contains two distinctive types of zircon. The most prominent occurs as red prismatic crystals up to 1 cm long enclosed in basalt. The crystals show resorption, indicating they have been accidentally caught up in the basalt magma. This is supported by dating of the zircons. Fission-track methods gave a zircon

◊

3.38 Sapphires from the Blackwood district (largest is 7 mm).

age of 1.8 million years, which may be the age of the eruption. However, isotopic dating methods gave various ages up to 5 million years, indicating the time when the crystals formed. An important specimen found at the quarry in 1994 shows two perfect, dark-red zircon crystals enclosed in alkali feldspar. This may be the remains of a small xenolith of syenite, a likely source rock for these zircons.

The other type of zircon found at the Ridge Road quarry, and at Leonards Hill itself, occurs as prismatic red crystals which may be up to 1 cm long but only 1 mm thick. These often contain unusual inclusions or tubes running along the crystal axis. Both types of zircon occur in the alluvial deposits in Stony Creek, which runs along the edge of the lava flow, and in Sailors Creek.

Blackwood–Trentham

This goldfield straddles the Great Dividing Range, about 70 km northwest of Melbourne. It consists partly of a series of shallow leads associated with the present-day streams which form the northern headwaters of the Lerderderg River and the southern headwaters of the Coliban River. The field extends for about 12 km between the towns of Blackwood in the south and Trentham in the north. The main discoveries of gold in creeks in the Blackwood area were in 1855, while rushes to creeks in the Trentham district took place in 1859 and 1862. The northern part of the field became known as the Blue Mountain diggings, after a prominent volcanic hill on the Divide midway between the two towns. The southern diggings sometimes went by the name of the Mount Blackwood field, after another prominent volcano about 10 km southeast of Blackwood. Mining of quartz reefs followed the depletion of the alluvial deposits, particularly around Blackwood (Ferguson, 1906).

Sapphires and zircons from the Blue Mountain diggings are amongst the earliest records of the gemstones in the State. Bleasdale, Ulrich and Smyth all referred to Blue Mountain sapphires in 1866, and sketches of zircon crystals accompanied Ulrich's essay. Specimens of blue sapphires and pink rubies, with zircons in 'titaniferous sands' were given to the Mines Department by the Mining Registrar, Mr R. H. Horne, prior to 1866. These samples are now in Museum Victoria's collections. Sapphires from the Blackwood area were first recorded in 1870 (Smyth, 1870). In 1906, a general description was provided by W. H. Ferguson, who stated that blue and greenish-blue sapphires, red zircons and black pleonaste were abundant in the alluvial deposits in the region. In the vicinity of Blackwood, the Lerderderg River at O'Briens Crossing, and tributaries such as Yankee Creek, have been popular with gem fossickers in more recent times. Many of these localities are now in a State park.

Blackwood–Trentham sapphires and rubies show a wide range of beautiful colours. Royal blue, bluish-green, purplish-blue, pale-lilac, and rare brownish-purple and yellow stones occur in samples collected by Ferguson and Dunn. They are commonly transparent to translucent, with some of the blues of gem quality and others showing faint colour banding. Dull translucent to opaque greyish-blue stones are also present. The fragments include rounded hexagonal 'barrels' and tapering 'points', flattened cleavage chips, bean shapes and irregular forms. The size ranges from 2 to 8 mm across.

Recently, weathered dykes exposed in road cuttings in the Blackwood district have been shown to contain variable proportions of zircons and sapphires.

3.39 Sapphires and rubies from Blue Mountain diggings, near Trentham (largest is 4 mm).

Blue Creek – Garlicks Lead

North of Blue Mountain and about 3 km west of Trentham, the upstream part of the shallow lead along Blue Creek runs parallel to a deep lead beneath a patch of high-level basalt. This deep lead was worked from 1863 and the township established there was named Garlicks Lead after the publican. The town later became known as Blue Mount and is now called Newbury. Fine blue and green sapphires, as well as a diverse range of zircons, can be found in Blue Creek to the north of the township. Some of the sapphires and larger rounded yellowish-brown to scarlet zircons have been recycled from the Garlicks Lead deposits into the creek. Rubies from the early Blue Mountain diggings include some highly lustrous, transparent lolly-pink stones. While very rare, similar rubies occur in samples collected in the 1990s in gold-mining concentrates from Blue Creek. Also present are dark purplish-red, bright pink and lilac stones, mostly rough fragments between 2 and 5 mm across. Slightly larger blue, green and rare yellowish-brown, transparent but flawed sapphires are also present, as rough fragments and rounded crystal points.

The alluvial deposits along Blue Creek abound in brilliant scarlet zircon crystals which show moderate resorption. Similar zircons have been found in decomposed basalt at Newbury, and in fresh basalt from the Garlicks Lead flow. These zircons have a fission-track age of 4.7 million years. Also abundant are sharply crystallised zircons, between 1 and 2 mm long, which have weathered out of the trachyte erupted from Blue Mountain 6 million years ago. These crystals are usually pale orange, but a few show blue-green colours, sometimes parti-coloured with orange. The cause of the striking colour, which may approach vivid blue, is a mystery. Another common variety

◊

3.40 Sapphires from Garlicks Lead (largest is 1.1 cm).

of zircon occurs as opaque crystals with two prismatic forms. These have fission-track ages of about 6.5 million years and, due to their abundance and lack of abrasion, probably have a nearby source. The large rounded brownish to scarlet zircons from the deep lead are about 50 million years old, based on fission-track dating (Hollis and Sutherland, 1989).

Guildford–Vaughan–Loddon River

The Loddon River closely follows its former valley as it flows northwards then westwards. Elongated patches of high-level basalt, overlying Cenozoic alluvial gravels, mark the former valley. These basalt cappings widen west of Guildford, forming a prominent feature known as the Guildford Plateau. The alluvial gold deposits below the basalt on the plateau were exploited by several large mines, after the first tunnels were put in during the late 1850s. Several companies were operating at Guildford during the 1930s (Mulvaney, 1939), and alluvial deposits in the Loddon River nearby were dredged for gold. Further upstream, Vaughan was the centre of the Fryers Creek Goldfield, named for a tributary of the Loddon River.

The sapphires from the Guildford Plateau alluvial deposits were the first to be described in some detail in Victoria, after a miner showed samples to George Ulrich in 1860. Ulrich noted blue stones, some with crystal faces and a fine colour, several green stones and a rose-coloured 'ruby', as well as transparent red zircons. Sapphires were recorded from Vaughan and Fryers Creek, and zircons from Kangaroo, near Tarilta (Ulrich 1867). Zircons and blue sapphires from the junction of the Loddon River and Boundary Creek were sent to the National Museum of Victoria in the 1860s. It is still possible to collect sapphires and zircons by panning gravels in the Loddon River, for example at Vaughan Springs.

Castlemaine

Zircons with a brilliant lustre resembling diamonds are registered in the Museum Victoria collections from Diamond Gully, near Castlemaine. Diamond Gully is a small tributary of Campbells Creek a few kilometres southwest of the town. The zircon crystals are highly rounded, transparent, colourless to pinkish-

◊

3.41 Zircons from Diamond Gully, Castlemaine (largest is 4 mm).

red, and up to 3 mm long. The Diamond Gully zircons have fission-track ages of 6 million years. Their source has not been found, but may be patches of Cenozoic gravels on the adjacent hills. Similar brilliant zircons were recorded from the Hard Hills, Campbells Creek by Ulrich (1867).

Lyonville

Several different types of zircon occur in the Loddon River north from Lyonville, a township about 6 km west of Trentham. They show very little abrasion, indicating they are from nearby sources, such as the 5.94-million-year-old trachyte flows forming Babbingtons Hill. Another unknown source provides dark-grey to orange-brown pyramidal crystals. Fine pale-blue sapphires occur rarely.

Shepherds Hut Creek

Another trachyte complex forms Spring Hill, about 9 km north of Trentham. Streams draining the northern side of Spring Hill are rich in ilmenite, and also carry a fine array of zircon crystals and rare clear-blue sapphires. Most of the zircons are brownish-orange, acicular crystals, and show a wide range of crystal forms.

3.42 *(opposite)* Zircon crystals from Lyonville showing range in colour and shape (largest is 3 mm). (Collection of Julian Hollis).

Allens Creek

The bed of Allens Creek at Allens Creek Road, a few kilometres south of North Blackwood, contains superb zircon crystals with forms more typical of granitic rocks. The crystals are up to 4 mm long and show hour-glass zoning from orange to colourless. Associated rare sapphires are mostly blue with a greenish-black alteration rind, possibly consisting of hercynite. Yellow-orange, euhedral crystals of monazite up to several millimetres long also occur. The minerals appear to be derived from a weathered breccia pipe, which would account for their excellent preservation.

Goodfellows Hill

Red soils on this low basalt lava dome, 6 km northeast of Trentham, contain abundant blue sapphires and orange zircons. Streams draining to the north contain concentrates of small zircons with sapphires and euhedral pink garnets which are probably derived from nearby Permian glacial sediments. Similar zircons are abundant on Cranneys Hill, a volcanic centre southeast of Trentham.

Campaspe River

There are several early reports of zircons from the Campaspe River. One locality for red zircons and rare blue sapphires is downstream of the former Mitchells diggings, in the Barfold Gorge near Redesdale (Selwyn *et al.*, 1868). Beautiful green crystals of zircon from the Campaspe River at Axedale were noted by Newbery in his 1870 Laboratory Report. Rounded reddish-brown zircons, with rare sapphires, can be found in the Campaspe River near Ashbourne, about 10 km southwest of Woodend.

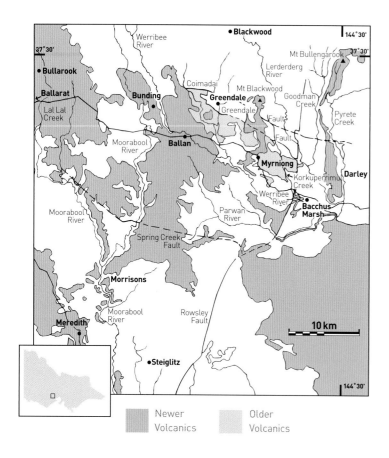

BALLAN REGION

This region is centred about 50 km WNW of Melbourne, immediately south of the West-Central Highlands (described above). It is defined geologically as a graben, a down-faulted block between the Greendale–Coimadai Faults in the north and the Spring Creek Fault to the south. The basement rocks are folded Ordovician siltstones and slates, intruded by granite of Devonian age and overlain by sediments deposited by Permian glaciers. In turn, these are partly covered by old lava flows of the Older Volcanics, then a layer of Cenozoic clays, sands and gravels, with some coal beds. Much of the floor of the graben has been covered by younger lava flows (of the Newer Volcanics) from volcanoes, such as Mount Blackwood, on the north rim. The down-faulting has protected the Permian and younger rocks from erosion; in particular the large volume of Older Volcanics preserved in the northern portion of the graben is significant as a source of zircons and sapphires. Zircon ages of between 50 and 77 million years have been obtained, similar to the age range shown by the volcanic rocks (Sutherland, 1996).

Although some alluvial gold was found in the region it was not particularly rich. Sapphire and zircon localities in and around the graben are poorly documented, but are amongst the most prolific sources in Victoria.

Pyrete (Coimadai) Creek

Bleasdale (1876) described a perfect translucent yellow bipyramidal crystal, with blue banding, from Coimadai, 'near Mt Blackwood'. He also reported rubies and rolled, waterworn green sapphires from 'Coimadai Creek and Gully'. Now named Pyrete Creek, this is the eastern twin of the lateral streams formed along the prominent ridge of lava from the Mount Bullengarook volcano to the north.

3.43 *(above)* Simplified geological map of the Ballan region.

3.44 *(opposite, top)* Faceted sapphires from Morrisons (largest is 5 mm) (Private collection).

3.45 *(opposite, below)* Sapphires from Morrisons (largest is 8 mm).

A deep lead which was worked for gold occurs under the flow (Officer and Hogg, 1897; Stirling, 1899d) and is probably the source of the sapphires described by Bleasdale.

Myrniong

Patches of high-level gravels trending WNW from Myrniong towards Bunding are the remains of Cenozoic stream channels cut into the underlying Older Volcanics. The gravels contain rich concentrations of sapphires and zircons, similar to those found at Morrisons (see below). The sapphires are mostly fine blues, with rare mauve to pink varieties. Zircons are mainly yellow to grey and opaque, but transparent, yellow and orange-red, gem quality stones up to 1.5 cm across also occur. Grains of black spinel, pyroxene, amphibole and ilmenite occur as well.

Greendale

Pot-holes along the course of Dale Creek, north of Greendale, yield abundant zircons, some of gem quality. They are similar to those from Myrniong and appear to have been recycled from former gravels which have been removed by erosion, north of the Greendale Fault. Sapphires are, however, very rare in the pot-hole concentrates.

Morrisons

Gold-bearing gravels beneath basalt near the township of Morrisons, 75 km west of Melbourne, have been known since 1861 (Flett, 1970). The deposits are exposed in the valley of the Moorabool River for about 4 km northwards from the town, and were worked by a series of adits (Dunn, 1907). According to Canavan (1988), the gravels form a sheet, rather than a true channel deposit or lead. They are possibly Late Miocene–Pliocene in age (Bolger, 1977). Small sapphires and zircons occur at the locality, which has been popular

◊

3.46 Moorabool River valley, near Morrisons.

with gem fossickers. The gem minerals are panned from gravels in the river bed, but are derived from the deposits beneath the basalt.

The sapphires occur as rough, waterworn fragments, cleavage chips and rounded crystal points and barrel shapes up to 12 mm but usually less than 5 mm across. Colours range from deep royal blue to watery shades of greenish and purplish-blue. Distinct green shades are rare. Some small barrels show irregular distribution of coloured and colourless zones. Smaller irregular to rounded grains, deep red,

lolly-pink and purplish in colour, also occur. These resemble the rubies from Blue Creek near Trentham. A range of zircons occurs with the sapphires. Large, well-rounded, fractured, translucent fragments to 1.4 cm occur with pale-brown, orange-brown and deep-red rounded fragments between 2 and 6 mm across. Some highly lustrous red, orange and yellow crystals are also present. Several types of garnet occur in the concentrates (see Chapter 6), along with abundant irregular, rounded crystal fragments of black spinel up to 1 cm across.

Other localities

Sapphires have also been recorded in this region from Steiglitz, firstly by Smyth in 1869, and as recently as the 1970s in gem club reports. Mount Blackwood (Bleasdale, 1876), Lal Lal Creek and Werribee Gorge also yield sapphires. Small blue sapphires, 1–2 mm across, accompany reddish-brown zircon crystals, fragments and rounded grains in Moorabool East Creek, west of Bunding. Abraded zircon fragments, some showing crystal forms, occur in the gravels of Korjamumnip Creek, northeast of Ballan. Up to 1 cm across, the zircons are commonly transparent and show colours including pale pink and yellow, orange, honey-brown and deep brownish-red. In the Werribee Gorge, zircon occurs as abraded red crystals up to 16 mm across, rounded transparent honey-brown grains and pale greyish-brown opaque fragments up to 7 mm. Zircons also occur in gravels in the Lerderderg River.

◊

3.47 *(left)* Zircons from the Werribee Gorge (largest is 1.2 cm).

3.48 *(below)* Faceted sapphire showing 'silk' and zircon (4 mm) from Morrisons (Private collection).

BALLARAT–CRESWICK REGION

Sapphires and zircons were reported from the Ballarat goldfields during the 1850s and 1860s, but usually without detailed descriptions or localities. Amongst the earliest collected specimens are rounded brownish zircons from the Yarrowee 'lead', the main stream flowing through the town. Deep red zircons (hyacinths) from Ballarat were displayed in rough and cut form in the Royal Society's 1865 exhibition. The gemstones came mainly from deep leads, which are covered by basalt flows to the west of Ballarat. Dumps from deep leads south and west of Ballarat were reported by Stone (1967) to be sources for sapphire and zircon, as well as topaz, garnet and quartz. The deposits are complex, with up to five distinct ages of gold-bearing sediments present, the oldest possibly Cretaceous. There are patches of early Cenozoic quartz–kaolinite gravels preserved in the region which are also a source of gemstones.

Adekate Creek

Clarkes Hill is one of several Newer Volcanic centres northeast of Ballarat. The stream system draining the western side of Clarkes Hill flows towards the deep leads at Creswick. Adekate Creek, upstream from the bridge on the Dean–Creswick Road, is a rich source of remarkable acicular zircon crystals. These have peculiar axial tubes with branching tubules at right angles, resembling hieroglyphics. Associated minerals are ilmenite, rare blue sapphire and perfect crystals of blue anatase.

Berringa

Sapphires and zircons occur in gravels in the bed of Mount Misery Creek, near Berringa, about 25 km SSW of Ballarat. The sapphires are waterworn, 2–4 mm across, and include blue (some transparent), greenish, intense pink and 'star' varieties. Zircons are similar in degree of abrasion and size to the sapphires and show shades of red. Black spinel crystals are also present.

Berry Lead

Sub-basaltic leads extending for about 12 km north of Creswick were mined for rich gold from the 1870s to about 1913. The main Berry Lead joins the Moolort Lead to the north, which in turn joins the lead along the present course of the Loddon River near Guildford. Mullock dumps on the Berry–Moorlort Lead near Creswick have been reported as sources for gemstones such as sapphire, ruby, zircon, topaz and garnet (Stone, 1967; King, 1985), but no descriptions have been published and the whereabouts of any specimens are unknown.

Buninyong

Older Cenozoic gravels (known as the White Hills Gravels), exposed in the sand and gravel pits west of Buninyong, contain rounded and abraded amber zircons and rare blue gem sapphires. The gemstones are similar to those from the Ballan region and are associated with spinel, ilmenite, and rare native lead and copper.

Bullarook

Rounded pale-brown to reddish-brown zircon grains to 8 mm have been found with younger elongated red crystals in the Bullarook Lead, south of Newlyn. This lead is an upper tributary of the Berry Lead. The young red crystals may be derived from the Leonards Hill volcano to the east.

Lake Learmonth

The sandy eastern shores of this maar volcano, 20 km NW of Ballarat, contain sparse zircon with rare sapphire, orange garnet and pyroxene. Some of the gemstones may be recycled from Cenozoic leads. Similar but rarer suites occur in a maar at Callender Bay, in the northwest of Lake Burrumbeet.

Talbot

Fine blue sapphires were reported from the Talbot and Mount Greenock gold diggings by Bleasdale in 1866. These workings were on a section of the Greenock–Chalks Lead, between Clunes and Maryborough. No detailed descriptions or specimens are known.

SOUTHWESTERN VICTORIA
Lake Bullenmerri

There is an isolated and restricted occurrence of large zircon crystals in a thin volcanic ash horizon on the eastern shoreline of Lake Bullenmerri near Camperdown (McMahon and Hollis, 1983). The crystals range from greenish to honey-coloured and brownish-red and are mostly pyramidal forms. They are commonly between 4 and 8 mm across, but may reach 1.7 cm. Although very lustrous, the crystals are also intensively cracked, possibly as a result of rapid cooling during eruption of the ash. They occur with abundant ilmenite.

Colac

Sapphires were reported to occur in the workings of the Colac Prospecting Company in 1871 (Smyth, 1871). These may have been found with gold in the Cenozoic gravels capping low hills to the south of the town (Flett, 1970). Sapphires were also recorded from a tributary of the Gellibrand River near Cape Otway by Selwyn *et al.* (1868).

3.49 (*above*) Collecting zircon crystals at Lake Bullenmerri (1993).

3.50 (*right*) Zircon crystals from Lake Bullenmerri (largest is 1.1 cm).

◇

3.51 Zircons from Carapooee (largest is 7 mm).

ARARAT–ST ARNAUD REGION
Carapooee

The diamond found in the Cenozoic gravels at Carapooee, 12 km SSE of St Arnaud (see Chapter 2), was accompanied by a variety of coloured sapphires and zircons. In hand-picked samples from the coarse concentrates, the sapphires are generally dull-looking crystal fragments between 3 and 8 mm across. The shapes are irregular, with flattened, waterworn fragments most common; no crystal faces have been observed. Despite a brownish iron-oxide patina, the fragments show a wide spectrum of colours, with shades of green showing yellow, blue and grey overtones predominating. The colours are watery, rather than intense. Blue stones are uncommon, whereas fragments showing purple shades, from pale amethyst to purplish-grey and brown, are conspicuous. A few pale to bright pink stones occur also. While most fragments are transparent, they are generally heavily flawed and unsuitable for faceting. However, it has been possible to select a small number of fragments to be cut into gemstones (Birch, 2008). Analyses of the full range of coloured sapphires suggest that they are derived mainly from rocks with metamorphic/ metasomatic character (Birch *et al.*, 2007). The main population of zircons from the coarse concentrates shows a limited colour range from pale yellowish-brown to orange-brown. However, like the sapphires, their overall colour, which is much greyer, is partly masked by a faint iron-oxide film. They are transparent, well-rounded fragments without any crystal forms evident, and with maximum dimensions between 3 and 7 mm. Dating by fission-track methods has determined their age as 67 million years. A much smaller population of transparent, lustrous, yellow to orange zircon crystals which show remnant faces is slightly older (74 million years). Associated heavy minerals are rutile,

black spinel, pseudorutile, andalusite, schorl
and anatase. Grains of opaque, pale to dark
blue corundum intergrown with muscovite,
which probably have a metamorphic origin,
also occur. Neither the distribution of the
transparent sapphires and zircons in the
gravels, nor their primary source, is known,
although they are probably derived from
basaltic volcanic rocks (Birch et al., 2007).

◇

3.52 Faceted sapphires from Carapooee
(largest is 3 mm).

3.53 (*above*) Sapphires from Carapooee
showing colour range (largest is 7 mm).

Other localities

Deep lead dumps south of Ararat are reported to yield sapphires and zircons, along with topaz, coloured quartz and garnet (Myatt, 1972; Stone, 1967). Transparent pale-brown to yellow zircons to 5 mm have been collected from near Rheola. However, there are no published descriptions of sapphire and zircon occurrences in this region. Various forms of metamorphic corundum occur in the high-level gravels near Amphitheatre (see Chapter 9).

WESTERN VICTORIA
Coleraine district

Waterworn blue and bluish-green sapphires up to about 5 mm across, with larger fragments of black spinel, have been found in Konong Wootong Creek, north of Coleraine, in western Victoria, although the exact location is unknown. The creek's headwaters are immediately south of those of Mathers Creek, which drains through Kongbool Station, where diamonds, sapphires and zircons were reported in 1895. A blue sapphire was found during exploration by CRA Exploration in Wennicott Creek (Rickards, 1991). These scattered occurrences suggest there may be a common source for the gem minerals, possibly related to small Older Volcanic plugs and lava flows north of Coleraine. Alternatively, the gemstones may have been recycled from early Cenozoic alluvial sediments or Permian glacial deposits in the region (Morand et al., 2003).

Origin of the sapphire–zircon suite

Sapphires and zircons are found along the eastern margin of Australia, from north Queensland to Tasmania. Their distribution coincides with basaltic lava fields, which range in age from early Cenozoic to nearly the present day, and overlaps the major diamond fields of New South Wales. This suggests the gemstones and basalts may be related in some way. The close association is often confirmed by the abundance of zircons and sapphires in soils developed directly from the old lava flows. Very rarely, isolated large crystals of zircon have been found in fresh basalt, for example at Ridge Road Quarry near Daylesford, near Lake Learmonth, and at Garlicks Lead near Trentham. Sapphires have not been found directly in basalt in Victoria, but examples are known from central Queensland and New England, New South Wales (Sutherland, 1996). Rich concentrations of sapphire have also been found in weathered volcaniclastic deposits in New South Wales. These deposits result from explosive eruptions early in the volcanic history of a region and may be the main source of alluvial sapphires (Oakes et al., 1996). No similar deposits have yet been found in Victoria, possibly because they are often weathered and difficult to recognise. However, the colour range and varied habits shown by the Victorian sapphires are typical of eastern Australian sapphires (Coldham, 1985), and suggest they have had similar origins and methods of emplacement (Birch et al., 2007; Birch 2008).

The ages determined on eastern Australian zircons suggest they mostly formed at the same time as the volcanic activity within each particular gemfield (Sutherland, 1996). Many of the gem zircons possibly crystallised in rocks rich in feldspar and nepheline, known as syenites, formed in the lower crust as a result of partial melting of the upper mantle. Later-formed basalt magma dissolved the nepheline and feldspar in these rocks, releasing the zircons and carrying them to the surface. Corundum may crystallise in different environments, such as metamorphic rocks, contaminated magmas and pegmatites, from the lower to the upper crust (Oakes et al., 1996; Sutherland et al., 1998; Sutherland and Fanning, 2001; Sutherland and Schwarz, 2001; Sutherland et al., 2009). Many of the sapphire and zircon crystals have been partially dissolved, indicating they were 'accidental' crystals or xenocrysts, caught up in and resorbed by the magmas which brought them to the surface. There is still no clear explanation for the variety of colours and habits of the gem minerals, other than that these events took place many times and involved different types of host rocks crystallising at different temperatures and pressures.

The results of this study of Victorian sapphires and zircons suggest there are some regional distinctions between the types of sapphires and, to a lesser extent, zircons. This may indicate there are distinct source rocks supplying sapphires for each region. On the other hand, some distinctive varieties of corundum, such as the chatoyant hair-brown stones and the bright pink rubies, occur across several regions. There is scope for considerably more research on the geological distribution, chemical composition, ages and source rocks of the Victorian sapphires and zircons.

◊

3.54 Faceted yellow-green Anakies sapphire set in ring. Sapphire mined at Bedfords Hill, Rubyvale, between 1915 and 1920 and faceted in 1960 (Private collection).

Sapphires, rubies & zircons in other Australian states

Gem corundum localities extend all the way from Tasmania to northern Queensland, following the distribution of Cenozoic basaltic volcanic rocks. Sapphires and rare rubies, along with zircons, are found in alluvial deposits in streams flowing off the basalts. The Anakie fields in central Queensland and the New England region in northern New South Wales are rich enough to have been mined for sapphires. Elsewhere, such as in the Ringarooma Valley in northeastern Tasmania, the gems are mainly of interest to collectors with an eye for faceting. The association of zircons and sapphires is very useful for scientists, because the ages of the zircon crystals can be determined by special techniques that allow correlation between the formation of the gems and periods of active volcanism.

The main location in Australia for gem-quality zircon crystals is far removed from the eastern Australian gemfields. Instead, it's a remote deposit in the Strangway Range, east of Alice Springs in the Northern Territory. Here, large zircon crystals are weathering out of a rock type unusual for Australia, known as carbonatite.

ANAKIE DISTRICT
The Anakie sapphire fields are nearly 400 km inland from Rockhampton, around the towns of Rubyvale and Sapphire. The sapphires are mostly mined from alluvial deposits, known as 'wash', in tributaries of the Nogoa River. Some of the most popular localities with fossickers include Retreat Creek, The Willows and Tomahawk Creek, where the 'shovel and sieve' method is used to mine the sapphires from shallow wash. The wash has been redeposited

after the erosion of sapphire-rich layers that were erupted not as lava flows but as explosive volcanic debris. Zircon dating suggests these eruptions occurred around 60 million years ago.

Blue, green, yellow and parti-coloured sapphires are found in the Anakie field. Colour as well as size varies in distribution throughout the field, with the best blue stones generally coming from Retreat Creek. Crystals tend to show rough hexagonal pyramid shapes, and basal cleavage fragments are also common. Most of the blue stones are exported to Thailand, where they are cut and often marketed as 'Thai' sapphires. Occasionally, exceptionally large gems have been recovered, such as the Black Star of Queensland (1156 carats) and the four stones carved into heads of US presidents (1713 to 2302 carats).

◊

3.55 Rough and faceted sapphires from the sapphires from the Anakies field, Queensland (largest cut stone is 10 mm across and 4.4 carats). (Lower rough and faceted stones from the Collection of the Australian Museum, Sydney).

BARRINGTON TOPS

Rubies show the same distribution as sapphires in eastern Australia but are much rarer. The richest deposits are on the Barrington Tops plateau, in the Upper Manning River district of New South Wales, where small-scale commercial alluvial mining operated for a few years from 2005. The rubies are purplish to pinkish red crystals, generally less than 6 mm across, but with some faceted stones exceeding 1 carat.

3.56 Rough and faceted pink sapphire and ruby from Barrington Tops, New South Wales. Cut rubies measure 2–4 mm across (0.15–0.34 carat) and are typical in size, colour and clarity for this area. (Collection of the Australian Museum, Sydney).

HARTS RANGE

Rubies with an entirely different origin to those in eastern Australia were found in the Harts Range, Northern Territory, in the late 1970s. A band of Precambrian gneiss near Mount Brady contained spectacular, deep-red hexagonal ruby crystals up to 5 cm across and 1 cm thick. Despite their intense colour, the rubies have mainly been used for cabochons, due to their opacity and internal fracturing.

MUD TANK BORE

Low hills a few kilometres south of Mud Tank bore, about 100 km northeast of Alice Springs, are formed in carbonatite, which is an unusual intrusive rock composed mainly of calcite. Large crystals of zircon, fluorapatite and magnetite (now altered to hematite, and called 'martite') have been released from the carbonatite as it weathered, and are now scattered over the surface, where they are sought after by gem fossickers. Zircon occurs as well-formed prismatic crystals up to 25 cm high, but is much more common as irregular fragments, in colours ranging from pale brownish grey to deep reddish brown. Attractive gemstones have been faceted from transparent material.

3.57 Ruby crystal (2 cm across) from Harts Ranges.

◊

3.58 *(above)* Zircon crystals
(4.5 cm) from Mud Tank bore,
Northern Territory.

3.59 *(right)* Faceted zircons
(largest is 9 mm) from Mud Tank
bore, Northern Territory.

Olivine &
Anorthoclase

◇◇◇◇◇◇◇◇◇◇◇◇◇◇◇◇◇◇◇◇

The Cenozoic basaltic lava flows, which form the volcanic plains of the western district of Victoria, are not generally thought of as a source of gem minerals. However, fragments of the silicate minerals olivine, anorthoclase, plagioclase, pyroxene and hornblende found at some volcanoes offer potential for unusual gemstones. The minerals may occur in xenoliths, which are dense crystalline rocks brought up from the Earth's mantle and lower crust by the magma, or may form large crystals known as megacrysts.

Xenoliths and megacrysts of these minerals are also present in some of the older Cenozoic basaltic lava flows preserved in the Ballan region, on Phillip Island and at Mount Lookout in Gippsland. Fragments of olivine and pyroxene are sometimes found with sapphires and zircons in streams draining regions where these lavas occur.

The xenoliths and megacrysts provide important evidence for the origin of the lavas and the types of rocks forming the deep crust and upper mantle below Victoria. The Victorian suite is one of the most diverse in the world and has been extensively studied (Irving, 1974).

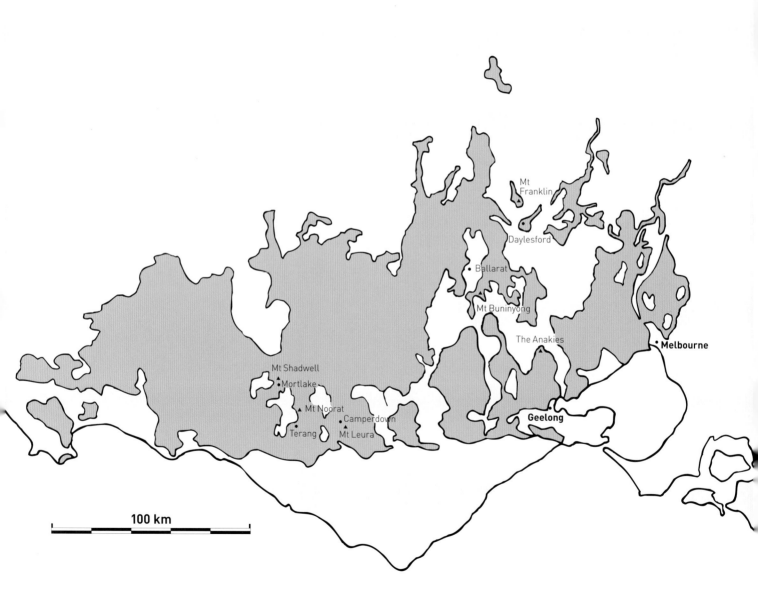

Mt Franklin

Mt
Franklin

Daylesford

Ballarat

Mt Buninyong

The Anakies

Melbourne

Mt Shadwell

Mortlake

Mt Noorat

Camperdown

Terang

Mt Leura

Geelong

100 km

OLIVINE

Olivine is very common as small crystals in the volcanic rocks. However, it is more conspicuous in xenoliths found mainly at scoria volcanoes, often as cores of volcanic bombs, and as pebbles on the shores of some volcanic lakes. There are several types of xenoliths, with the most common being lherzolite, a rock type composed mainly of grains of pale green olivine. The olivine contains more magnesium than iron and is more correctly called forsterite. Olivine in the xenoliths hardly ever shows well-formed crystal shapes. However, some grains are sufficiently large and transparent to be faceted as the gem variety known as peridot.

ANORTHOCLASE AND OLIGOCLASE

Crystal fragments up to 5 or 6 cm across of colourless to white anorthoclase and oligoclase, sodium-rich minerals of the feldspar family, can be found in scoria at some volcanoes. The crystals are often broken along cleavage planes, but may show some rounded surfaces caused by resorption in the magma. The origin of these megacrysts is uncertain, although they probably formed in magma on its way to the surface. Anorthoclase crystals are now being used for Ar-Ar dating to provide evidence for when eruptions occurred.

PYROXENE

Other common minerals forming megacrysts are glassy, dark-brown augite and enstatite, which are members of the pyroxene family. Another pyroxene mineral, diopside, occurs in many xenoliths as small transparent grains coloured an attractive deep green by traces of chromium. Most of these minerals are found occasionally as transparent pieces large enough to facet, although their use as gemstones is not widespread.

Early history of discovery

George Ulrich noted in 1867 that 'chrysolite' (the old name for olivine) was common as an essential mineral in the young basalts west of Melbourne. It was especially noticeable that masses of olivine grains occurred near craters such as Mount Franklin, the Anakies, and Warrion Hills. Ulrich also reported that large crystal fragments of oligoclase were found at the first two localities and also in the shore deposits at Lake Purrumbete. Early chemical analyses revealed both oligoclase and anorthoclase; for example, oligoclase occurred at Mount Anakie (Howitt, 1898) and anorthoclase at Mount Noorat (Grayson and Mahony, 1910). However, anorthoclase is the more common, especially at Mount Franklin (Mahoney, 1928). The volcanic history and some of the minerals found in the scoria at Mount Shadwell near Mortlake were first described by Jutson and Chapman in 1905.

4.1 Anorthoclase crystal in scoria and faceted gemstone (7.5 mm long) from Mt Franklin.

4.2 *(opposite)* Western Victoria showing lava distribution and localities in text.

Localities

Mount Franklin

Mount Franklin is one of Victoria's most important megacryst localities. It forms a prominent volcanic cone 9 km north of Daylesford. The volcano is about 185 m high and has a deep crater with a narrow breach in its rim on the southeastern side. The cone consists of scoria which contains pieces of olivine and megacrysts of anorthoclase and augite up to 11 cm across (Beasley, 1970; Coulson, 1954). Some of the anorthoclase crystals contain transparent portions from which small gemstones have been cut.

The Anakies

The Anakies are a group of three scoria cones and a small maar, about 30 km north of Geelong. The eruption points are aligned NW–SE, with the western and highest cone, reaching 170 m, forming Mount Anakie. The eastern and central cones are extensively quarried for scoria. An important suite of olivine-bearing xenoliths occurs in the scoria of the western and eastern cones. Anorthoclase and black amphibole megacrysts are abundant at the eastern cone, with pyroxene more common on Mount Anakie.

Mount Noorat

The Mount Noorat scoria volcano is 6 km north of Terang. It is significant both for its 150-m-deep, circular crater and for the abundant xenoliths and megacrysts found in the scoria, which has been quarried on the western side of the volcano. Lherzolite is the main rock type, while augite and anorthoclase are the most abundant megacrysts.

4.3 Mortlake road sign.

4.4 Scoria quarry, Mt Shadwell (2012).

Mount Shadwell

Mount Shadwell is the highest of
a group of scoria cones 2 km north
of the town of Mortlake. The scoria
contains many different xenoliths,
as well as megacrysts of augite and
anorthoclase. Some of the xenoliths
contain very coarse patches of
transparent, grass-green olivine
grains up to several centimetres
across, from which small gemstones

4.5 *(below)* Faceted olivine
(peridot) from Mt Shadwell
(largest stone is 8 mm long).

4.6 *(right)* Faceted chromian
diopside from Mt Shadwell
(larger stone is 6 mm across;
0.92 ct).

may be cut (Beasley, 1966). Some of the olivine contains more iron, giving gemstones with orange or brown shades. Deep-green chromian dropside grains have also been faceted. Mortlake has proclaimed itself to be the 'Australia's Olivine Capital' and fossickers are permitted to visit the scoria quarries on Mount Shadwell by prior arrangement with the manager.

Mount Leura

This volcanic complex on the eastern outskirts of Camperdown consists of a central scoria cone surrounded by smaller cones and a large maar crater and tuff ring. Scoria and tuff quarries operate on the northern edge of the volcano. Mount Leura is one of the most prolific sites for xenoliths, especially lherzolite, and megacrysts of anorthoclase are also common.

Mount Buninyong

This is a very prominent breached scoria cone, over 200 m high and with a 75-m-deep crater, about 13 km southeast of Ballarat. It has been a popular destination for gem fossickers collecting olivine-rich xenoliths in the scoria.

Olivine & anorthoclase in other Australian states

Gem feldspar crystals are found in many of the basaltic volcanic rocks in eastern Australia. They are mainly the sodium-rich feldspar anorthoclase, forming rounded glassy megacrysts embedded in scoria. Transparent crystals ranging from labradorite to andesine, intermediate members of the plagioclase series of feldspars, are much rarer, and are best known from the Hogarth Range in northern New South Wales and near Springsure in central Queensland.

Gem-quality olivine (peridot) crystals have also been found at several localities in these basaltic volcanic regions. The best crystals for faceting come from localities in Queensland.

4.8 *(below)* Gem-quality labradorite crystal (3 cm long) from the Hogarth Range, New South Wales.

4.7 *(above)* Faceted anorthoclase from Mt Shadwell (larger stone is 14 mm across; 7.4 ct).

Hogarth Range

Colourless to pale-yellow crystals
and fragments of labradorite–
andesine up to several centimetres
across occur associated with
Cenozoic (Miocene) basalt at a
locality in the Hogarth Range, near
Mallanganee, 40 km southwest
of Casino in northern New South
Wales. Fossickers find faceting-
quality crystals in soil formed by
the weathering of the basalt.

Moonstone Hill

This scoria cone is a designated
fossicking area in the Blackbraes
National Park, near Chudleigh,
some 130 km north of Hughenden,
Queensland. Crystals of colourless,
faceting-grade anorthoclase,
some of which display the typical
bluish adularescence or schiller
of 'moonstone', are abundant
in soil on the hill. Fragments
of gem-quality olivine (peridot)
occur with the feldspar.

(top to bottom)

4.9 Faceted labradorite (5.0 and
4.0 carats; 2.75 carats) from the
Hogarth Range, New South Wales.

4.10 Anorthoclase crystals (4 cm across)
from near Hughenden, Queensland.

4.11 Faceted peridot (1.25 and 0.9 carats)
(6 mm) from Springsure, Queensland.

Topaz, Tourmaline & Beryl

"I may venture to say that there is not in the world a stone fit for brooches, of size, and fire, and lustre, and suited to both day and candlelight, equal to some of the blue topazes which I have known to be found in Victoria."

REV. JOHN BLEASDALE, 1867

◇◇◇◇◇◇◇◇◇◇◇◇◇◇◇◇◇◇

Topaz, beryl and members of the tourmaline family are silicate minerals well known for their colourful gem crystals and found in many locations around the world. The gem minerals are most commonly found in cavities within granite, especially in coarse patches or veins of pegmatite. The crystals grow from solutions containing elements such as fluorine, boron and beryllium, which may become concentrated in magma as it cools and crystallises to form granite.

Despite the abundance of granite in Victoria, only schorl — the dark-brown to black member of the tourmaline family — could be said to be common and widespread. Coloured gem species of tourmaline, such as elbaite, are extremely rare. Topaz crystals have been found *in situ* in granite at only a few localities, although waterworn fragments are quite widespread in alluvial deposits which have had granite as a source rock. Beryl is the rarest of these minerals in Victoria, with only a few isolated occurrences of crystals in granite and alluvial deposits.

In this chapter, the main localities are grouped on a regional basis. Because it is so widespread, only the most important localities for tourmaline are described.

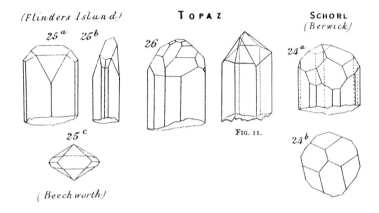

(Flinders Island)

25^a 25^b

25^c

(Beechworth)

TOPAZ

26

FIG. II.

SCHORL
(Berwick)

24^a

24^b

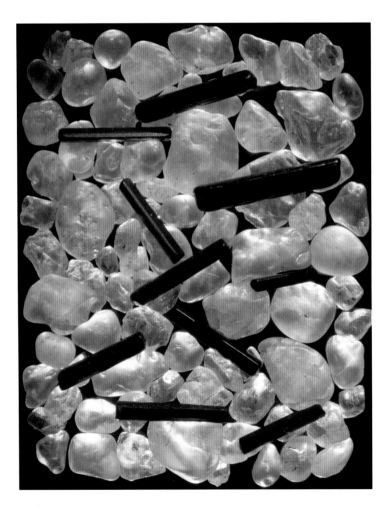

Early history of discovery

TOPAZ

The early reports of topaz all refer to waterworn fragments and crystals found in alluvial deposits in various goldfields. The Ovens District, centred on the Woolshed Valley, yielded abundant colourless to pale blue stones in the drifts. However the best quality pieces came from deposits in the leads at Dunolly, Talbot and Mount Greenock during the 1860s (Bleasdale, 1865, 1866; Ulrich, 1867). Several large topaz crystals were found in pebble wash from the Bradford Lead near Maldon prior to 1870 (Ulrich, 1870). The fine topaz crystals found on Flinders Island in Bass Strait were also reported in the 1860s. While discoveries of topaz crystals in alluvial gold deposits tapered off during the late 1870s, many were found during later tin mining in creeks in parts of Gippsland. These included small crystals from the Pakenham district that were examined and sketched by Charles Anderson, at the Australian Museum in Sydney, in 1908.

Many of the Victorian topaz pieces were of faceting quality and some were featured in the Royal Society's exhibition of 1865, as well as the Colonial and International exhibitions held in 1861 and 1872. During the 1870s several hundred crystals were faceted by the Melbourne jewellers Spink and Son.

TOURMALINE

Like topaz, black tourmaline was amongst the first minerals identified in the alluvial gold deposits, especially as water-rounded crystals in the Ovens District. However it was also recognised in granites, including those near Maldon and at Mount Alexander. Lustrous black crystals from 'Upper Goulburn' were amongst the earliest specimens collected and were featured in the Melbourne Exhibition of 1854. Another notable early discovery was of aggregates of black tourmaline in granite near Beechworth (Dunn, 1871). Prismatic crystals were found near Baynton, east of Kyneton, during mapping by the Geological Survey in the early 1860s (Selwyn, *et al.*, 1868). Most of the black tourmaline crystals are classified as schorl. Coloured varieties were rarely encountered. Red 'rubellite' crystals in quartz from Tarrengower (Maldon) described by Bleasdale in 1868 were identified later as garnets by Ulrich (1870). Green tourmaline crystals were reported from granite near Wangaratta (Bleasdale, 1865), in the Yarra River (Ulrich, 1870) and near Pakenham (Atkinson, 1897).

BERYL
There were a few early reports of emerald, the bright green variety of beryl, being found in the gold drifts (e.g. Becker, 1857). Emerald has also been reported from deep leads in the Ararat and Ballarat regions (Stone, 1967; King, 1977). These records are unlikely to be authentic, with green sapphires, or possibly chromium-bearing diopside, being misidentified as emerald.

5.1 *(opposite, top)* Sketches of topaz and tourmaline crystals (Ulrich, 1866).

5.2 *(opposite, below)* Waterworn topaz and schorl from Beechworth district (largest schorl is 1.8 cm long).

Localities

NORTHEASTERN VICTORIA
Ovens District
Topaz and tourmaline are common in the alluvial deposits of the Eldorado Lead system in the Beechworth region (see Chapters 2 & 3). Early reports described the topaz as pale blue or colourless, occasionally with a yellowish or pinkish tinge (Dunn, 1871; Bleasdale, 1876). Most of the fragments were well rounded and bean-sized, with some larger pieces up to 2.5 cm on edge. Dunn noted that perfect, sharp-edged crystals were found near Eldorado and Sebastopol, sometimes as a 'topaz-sand', suggesting nearby sources within the surrounding granite. Museum Victoria specimens of topaz collected by Dunn from many of the creeks show a mixture of shapes, including globules, ellipsoids, flattened cleavage plates and irregular fragments, generally less than 1 cm across. Most grains are transparent but flawed, and only small stones could be faceted from them. Waterworn topaz pieces were also found in the Lancashire and Indigo Leads, and as far north as the Great Southern gold mine on the Prentice Lead, near Rutherglen.

Waterworn topaz can still be collected in Reedy Creek along the Woolshed Valley and near Eldorado. During the 1970s, topaz crystal fragments several cm across could be collected in gravels at an alluvial tin mine on Clear Creek, near Byawatha, north of Eldorado.

Abraded crystals of black schorl, from slender to stumpy prisms up to 2.5 cm long, commonly accompany the topaz. The main schorl occurrence was in a clay-filled cavity about 1 metre across in granite exposed in Reids Creek during the 1860s. According to Dunn (1871), the cavity contained attractive compact sprays of black needle-like crystals in masses up to nearly 5 kg, accompanied by quartz crystals 20 cm long.

Beryl is much rarer, restricted to only a few recorded crystals. An unusual pale-straw-coloured crystal, 2.5 cm long and 0.5 cm thick, found in the wash of Pilot Creek in 1877, was tentatively identified as beryl (Couchman, 1878). Specimens preserved in museum collections include a 4-mm colourless crystal from Reids Creek and a broken, 1.5-cm, pale bluish-green prism.

Source of the gem minerals
Most of the topaz, tourmaline and beryl have probably been weathered out of the granite of the Mount Pilot Range. Small colourless topaz crystals have been found accompanying cassiterite and quartz at several of the small tin deposits in the granite (Hensleys lode, Gimbletts lode). Several dykes containing small topaz crystals intrude the granite north of Eldorado (Birch, 1984).

Other localities

Fractured black schorl crystals up to 2 cm thick and 15 cm long occur in massive white quartz near Barnawatha. Similar crystals occur in quartz at Days Creek, near Omeo, and with fluorapatite in quartz veins near Tallangatta.

Topaz occurs as an accessory mineral at the Womobi tungsten mine at Thologolong near Granya, where it forms opaque, greyish green masses and rough crystals a few cm across in quartz. It has also been recorded at the Walwa tin mine.

Several of the tin-bearing pegmatite bodies on the eastern slopes of Mount Wills, in northeastern Victoria, contain small amounts of bluish, greyish green and pinkish tourmaline crystals. These appear to have compositions in the series between schorl and elbaite.

EAST-CENTRAL VICTORIA
Strathbogie Ranges

The Strathbogie Ranges are a picturesque plateau region covering about 1500 square km, with its southwestern rim about 80 km northeast of Melbourne. The highlands were not settled until the 1880s, and even today there are only small scattered farming communities. The ranges are composed mainly of granite which formed during the late Devonian period. Waterworn crystal fragments of topaz and tourmaline, as well as varieties of quartz, can be found

in stream beds in the ranges. In a few places crystals may be collected in weathered granite. Local lapidary clubs regularly visit several of the localities but these are all on private land and require permission for entry. Very little published information is available.

The earliest reports of topaz and tourmaline localities were very vague, such as 'Strathbogie Ranges' and 'Molesworth' (Atkinson, 1897). In wash in Tallangalook Creek, Dunn (1917) reported small rounded topaz pieces with garnet, cassiterite and gold. Waterworn topaz fragments were also found in Hellhole Creek, a tributary of Tallangalook Creek.

In the mid-1990s, attractive transparent blue topaz crystals were found near Mount Wombat, the highest point on the ranges, about 14 km southeast of Euroa. The crystals occur in clayey gravel in the bed of Wombat Creek. Most are less than 3 cm across and are wedge-shaped due to the development of two prominent prism faces in the form {021}, but are usually modified by smaller faces of other prism and dipyramidal forms and a pinacoid. This habit is typical of topaz crystals from late-stage pegmatites (Menzies, 1995). The crystals range from colourless to pale and moderate

◊ **5.3** Scene in Strathbogie Ranges (1997).

blue and many are sufficiently transparent to be facetable. Their sharp edges and lustrous faces suggest that their source, a pegmatite vein or cavity in the granite, is close by, possibly in the bed of the creek. Several specimens of topaz attached to smoky quartz crystals have been collected, but most of the topaz occurs as unattached crystals loose in the gravel.

Found with the topaz are sharp to waterworn prisms of tourmaline which are mostly black schorl. However several crystals up to 3.5 cm long, while appearing to be opaque, show a spectacular length-wise colour zonation from red to yellow and green. These crystals have been shown by analysis to belong to the series between the species elbaite and olenite. Other crystals showing intense greenish and yellowish brown colours in transmitted light have also been collected. A group of pale-blue beryl crystals (the variety aquamarine) about 6 cm long has been reported as being found at the topaz locality.

Black tourmaline crystal fragments up to 5 cm across can be found in Brankeet Creek north of Bonnie Doon, in County Creek north of Ruffy, in Stewarts Creek and in

◊ **5.4** *(above)* Topaz crystals (2 cm) on smoky quartz from Wombat Creek.

5.5 Simplified geological map of the Strathbogie Ranges region.

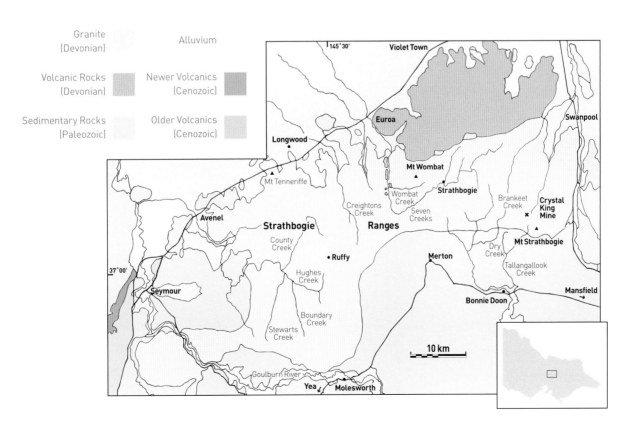

Granite (Devonian)

Alluvium

Volcanic Rocks (Devonian)

Newer Volcanics (Cenozoic)

Sedimentary Rocks (Paleozoic)

Older Volcanics (Cenozoic)

5.6 *(right)* Topaz crystal (1.8 cm across) from Wombat Creek.

5.7 *(top right)* Schorl crystal (3 cm long) from the Strathbogie Ranges.

5.8 *(far right)* Topaz crystal (2.2 cm) from Wombat Creek.

5.9 Topaz from Wombat Creek (4 cm crystal) and faceted stones, left and right) and Eldorado (centre).

places along Hughes Creek. A locality near Caveat, 10 km north of Molesworth, has been a popular collecting site for tourmaline and quartz crystals occurring in granitic soil. Thin, striated crystals of schorl up to 5 cm long, some with terminations, have been found with smoky quartz crystals in granite near Mount Tenneriffe, southwest of Longwood. Thick lustrous black schorl crystals up to 6 cm long and 3 cm thick, striated but rarely terminated, have been collected near Violet Town. A few well-terminated schorl crystals up to 3 cm across have been found at undisclosed locations near Swanpool.

Source of the topaz and tourmaline
The topaz and tourmaline found in stream gravels in the ranges are derived from pegmatites within the granite. In some cases, for example in Wombat Creek, pegmatites appear to be shedding crystals directly into the streams. At other localities, for example in County Creek and Stewarts Creek, broken and waterworn fragments of tourmaline, quartz varieties and, less commonly, topaz are derived from a layer of granitic gravel deposited on the granite bedrock. This layer may be up to several metres thick and represents an earlier period of erosion of the granite.

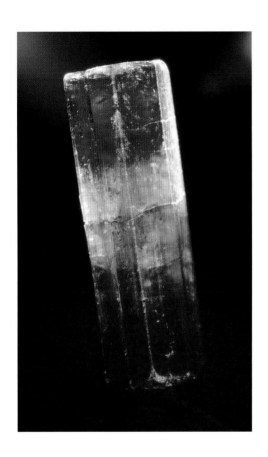

5.10 *(above)* Elbaite–olenite crystal (3.5 cm long) from Wombat Creek.

5.11 *(top right)* Elbaite–olenite crystals (2.3 cm long) from Wombat Creek.

5.12 *(right)* Elbaite–olenite crystal (3.5 cm long) from Wombat Creek.

Toombullup

Topaz is commonly found with zircon and sapphire in Webbs Creek (Kitson, 1896) and other streams such as Middle Creek and Stringybark Creek in the Toombullup goldfield (see Chapter 3). These are possibly the localities referred to as 'near Mansfield' by Bleasdale (1876). The topaz occurs as colourless to pale-blue rounded crystals and cleavage fragments up to about 6 mm across. Rounded black schorl crystals also occur. The sources of the topaz and schorl may be in the granite forming the eastern part of the Strathbogie Ranges.

Marysville district

Alluvial deposits and quartz reefs containing wolframite were mined in forested country about 8–10 km ESE of Marysville in the early 1900s. The reefs and the granodiorite host rock of the region often contained fibrous brown tourmaline, which formed veins and broken-up masses with quartz (Dunn, 1908). The mineral was first reported in the 1860s (Selwyn *et al.*, 1868). Analysis shows the brown tourmaline to be the species dravite. In the 1990s, transparent brown tourmaline crystals were found in a cutting in weathered Devonian rhyolite on Blue Range Road, northwest of Marysville. They appear to be derived from a pegmatite vein within the rhyolite.

Black Range ('Upper Goulburn')

Clusters of lustrous black schorl crystals, up to 2 cm long and often terminated, were found embedded in kaolinite and quartz in the 'Upper Goulburn' River region, during the early 1850s (Anon., 1854). Quartz crystals up to 30 cm long, with groups of schorl and feldspar attached to them, were also collected. The locality is probably somewhere in the Black Range, which is composed of several different types of granite forming the remote mountainous region between Healesville and Murrindindi (Howard, 1972).

Dandenong Ranges – Gembrook

Early reports listed Sassafras Creek and 'near Lilydale' as localities where topaz could be found (Atkinson, 1897). Small yellowish topaz fragments occur in William Wallace Creek near Gembrook. Rare topaz and green tourmaline crystal fragments have been panned from creek gravels near Belgrave.

5.13 Brown tourmaline (dravite) crystals from Marysville district (2 cm across).

5.14 *(opposite)* View towards Beenak from Gembrook.

GIPPSLAND
Beenak – Bunyip River –
Upper Tanjil River region

The Tynong Granite occupies much of the remote, mountainous country east of Gembrook, covering some 1100 square km. The granite has been the source for alluvial cassiterite found in many of the streams in the region. These include the Bunyip (formerly Buneep) and Latrobe rivers which flow south and southeast, and the Little Yarra River and McCraes Creek which flow north or northwest. Mining of the alluvial tin deposits began in the 1860s and continued intermittently until the early 1900s (Murray, 1891; Cochrane, 1971). The granite also supplied crystals of topaz, schorl and varieties of quartz which are found in many of the streams. For example, McCraes Creek near Beenak, and Rysons Creek, an east-flowing tributary of the Bunyip River north of Labertouche, were popular destinations for gem fossickers in the 1960s and 1970s. Further north, small topaz crystals were recorded with alluvial cassiterite and scheelite in Britannia Creek (Howitt, 1921).

Topaz is also widespread, if uncommon, in the granite country east of Powelltown. Topaz and tourmaline can be found with smoky quartz, amethyst and citrine in creeks near Starlings Gap, and in the Ada River and Federal Creek. Topaz has also been found with schorl and rock crystal in the Tarago River. Even further east, topaz and smoky quartz have been found on the Toorongo Plateau, at the head of the Tanjil River. In places along the Thompson and Tyers Rivers, topaz and zircon have been found in gravels.

In most of the alluvial deposits the topaz occurs as colourless to pale-blue, waterworn crystals and fragments, up to a few cm across. However at Rysons Creek, topaz pieces up to 20 g have been collected (Pauley, 1966).

The schorl occurs as waterworn black crystal fragments. No records of topaz or schorl within the granite itself are known.

South Gippsland

Topaz fragments occur in the alluvial gravels in the Franklin River tin field (Smyth, 1873), at Yanakie Landing (Murray, 1876) and in Stockyard Creek (Atkinson, 1897). Black schorl is very common on Wilsons Promontory, occurring as patches in granite and as prismatic crystals enclosed in quartz. Specific localities include Mount Singapore and the Mount Hunter tin deposits.

WEST-CENTRAL VICTORIA

Records of topaz and tourmaline in this region refer mainly to discoveries in scattered Cenozoic gravels which were mined for gold between the 1860s and 1890s.

◊ **5.15** Topaz crystals from the Bunyip River district (largest is 1.5 cm).

Talbot and Mount Greenock Lead

As well as sapphires, fine blue topaz was discovered in the Talbot and Mount Greenock diggings, between Clunes and Maryborough, during gold mining in the 1860s. Bleasdale (1865) referred to blue topaz pieces up to $1/4$ lb (110 gram) in weight which were flawless enough to be faceted. Ulrich (1870) described colourless to pale-blue, rounded grains, from pea-size to larger than a walnut. According to Bleasdale (1866), all the largest crystals went into private hands and no specimens appear to have been preserved in museum collections.

Majorca

Waterworn fragments of topaz and schorl were found in deep lead deposits in this goldfield near Maryborough. Specimens were obtained at depths of 160 ft (50 m) in the Robin Hood Claim and at 120 ft (36 m) in the United Kingdom Claim at nearby Gibraltar (Smyth, 1869). Some crystals were added to the collections of the Mining Department in the 1860s but have been lost.

Dunolly

Waterworn topaz crystals, from white to pale blue and up to 1.5 ounces (about 48 gram), were reported from the Dunolly goldfields. Ulrich (1867) described several crystals over an inch (2.5 cm) across, with a thin colourless outer zone around a limpid blue core. Specimens in the Museum Victoria collection are well-rounded to irregular fragments up to 11 mm across, from colourless to pale pink and pale blue.

Rheola

Topaz, garnet and tourmaline, together with a variety of quartz crystals, including cairngorm, citrine and amethyst, were found in cemented Early Cenozoic gravels in the Rheola goldfield. The main locations were Possum Hill and Humbug Hills, between Rheola and McIntyre (Dunn,

1890). Topaz crystals collected by Dunn from the Humbug Hills are rounded mainly blue stones, with one yellowish crystal, up to 1 cm across.

Bradford Lead

This is an isolated alluvial deposit preserved on granitic hills about 8 km north of the historic gold mining town of Maldon and just west of the former township of Bradford. Shafts up to 25 metres deep were sunk into the deposit during the 1860s but gold yields are unrecorded. According to Ulrich (1868, 1870), the Geological Survey party discovered sky-blue topaz in mullock from one of the shafts. A waterworn pebble (60 gram) and a worn crystal (9 gram) were collected and added to the National Museum of Victoria collection. A variety of other granitic minerals, such as garnet, tourmaline, smoky quartz and amethyst, were also found in the deposit (see Chapters 6 and 7). The mullock dumps were well turned over by gem collectors during the 1960s and 1970s, with further discoveries of topaz and smoky quartz crystals being reported.

5.16 Dumps on the Bradford Lead (2010).

Maldon

Patches of black schorl are common in the vicinity of Mount Tarrengower, the highest point in the granite ranges north and west of Maldon. Prismatic schorl crystals enclosed by white quartz are also widespread, for example in the Nuggetty Range.

Mount Bolton

Several crystals of the transparent yellow variety of beryl, known as 'heliodor', were found in a gravel pit in the Devonian granite

◊

5.17 Beryl crystal in smoky quartz crystal (10 cm high) from Mount Bolton (Private collection).

forming Mount Bolton, near Learmonth, about 25 km northwest of Ballarat, in 1988. They occurred in a cavity lined with smoky quartz crystals. The largest beryl crystal, about 6 cm long, was originally partially enclosed in a 10-cm-long smoky quartz crystal. Severe natural etching had removed the terminations and freed the crystal from its host. Other cavities have been found containing fine crystals of orthoclase, smoky quartz and amethyst, with rare garnet, but no other beryl crystals have been discovered.

Blackwood

In 1899, James Stirling reported that topaz occurred in a decomposed dyke between Darley and Blackwood. The dyke had been found during a survey traverse by Charles Brittlebank, and contained white to greenish topaz crystals up to 1.2 cm across. The position of the dyke is not known (the survey line is not documented) and no crystals have been preserved.

Baynton

During mapping of Quarter Sheet 51 SW by the Geological Survey in 1865, lustrous black tourmaline crystals were found in soil near Hells Corner, on Bayntons Station, about 25 km northeast of Kyneton. Terminated slender prisms and striated crystal sections up to 4 cm long were collected (Selwyn *et al.*, 1868). In transmitted light they show dark brown and dark blue zonation. The locality lies close to the northern margin of the Cobaw Granite, of late Devonian age, and the crystals are weathering out of pegmatite associated with the granite.

Other localities

Topaz has been recorded, without specific locality details or descriptions, from other alluvial deposits in the west-central region (Atkinson, 1897). These include Pleasant Creek (near

Stawell), Ballarat, Ararat, Moonambel (Mountain Creek), Bendigo and Castlemaine. A 'remarkable' topaz prism capped by milky quartz was found at Bealiba and exhibited in 1872 (Anon, 1872). Topaz has been found in the deep leads around Daylesford (for example in the Italian Hill Lead), in the Campaspe River at Redesdale and in the Werribee River in the Werribee Gorge. Topaz is also reported from the deep lead dumps in the Ararat and Ballarat districts (Stone, 1967).

Source of the topaz
The source rocks for the topaz in all the alluvial deposits are nearby granites. For example the head of the Talbot and Mount Greenock Lead rises in the granite forming Mount Beckworth and Mount Bolton to the south. The leads near Dunolly and Rheola are fed in part from surrounding granites, including that forming Mount Moliagul, while the Bradford Lead sits directly on the granite (the western part of the Harcourt Granodiorite) which forms Mount Tarrengower and the Nuggetty Ranges.

◊

5.18 *(right)* Pale blue beryl crystals (6 mm long) in the Lake Boga Granite.

NORTH-CENTRAL VICTORIA

Granite of Devonian age forming the Terricks Range, and isolated low hills near Pyramid Hill, Mount Hope, Mitiamo and Wycheproof, protrudes through the Cenozoic sediments of the Murray Basin in north-central Victoria. Quarries have operated on several of the granite outcrops, exposing many coarse pegmatite patches and cavities.

Lake Boga

The largest granite quarry in the region is about 10 km SSW of Lake Boga and has operated since 1924. The granite contains patches of pegmatite surrounding irregular cavities which may reach up to one metre across. These are lined with fine crystals of smoky quartz, orthoclase, albite, muscovite and fluorapatite (Birch, 1977), and may also contain rare copper and uranium phosphate minerals (Birch, 1993, 2008). Patches of dark bluish to black schorl are common in the granite, especially in the upper parts of the quarry. Prismatic crystals up to 9 cm long have been found in cavities, but are usually strongly striated and only rarely show terminal faces. Dustings of bluish grey powdery schorl needles occur on smoky quartz crystals. A 20 cm-wide pegmatite cavity containing massive pale blue topaz, and some small euhedral crystals up to 5 mm across, was found in 1995. Beryl is very rare, with only a few small prismatic colourless or blue crystals being collected.

Pyramid Hill

Aggregates of lustrous, pale-blue topaz crystals up to 4 cm across occur rarely in the granite being quarried at Pyramid Hill. Cavities containing several transparent blue topaz crystals up to a few cm across are also known. Aggregates and crusts of black schorl crystals are common. A few specimens of needle-like tourmaline crystals, dark reddish brown in colour and up to 3 cm long, have been found with calcite in cavities.

Wycheproof

Pegmatite veins in a granite quarry at Wycheproof contain aggregates of dark greenish brown schorl crystals up to 10 cm long accompanying a suite of rare zirconium phosphate minerals (Birch, 1993).

Topaz, tourmaline & beryl in other Australian States

Australia is not generally well known for deposits of gem-quality topaz, tourmaline and beryl, but there are a number of localities which have yielded colourful transparent crystals suitable for faceting. In eastern Australia, gem topaz is mostly found associated with granite, often as waterworn crystals in alluvial deposits. Mount Surprise and the Stanthorpe district in Queensland, the Oban area in New England, and Flinders Island in Bass Strait, Tasmania, are the main sources. Green to pink tourmaline crystals of faceting quality occur in pegmatite on Kangaroo Island, South Australia. Gem-quality green and blue tourmaline crystals have recently been found in the Giles pegmatite, near Coolgardie. Superb crystals of brown dravite and black povondraite occur in schist on Yinnietharra Station in the Pilbara district of Western Australia, but these are not of gem quality. Beryl is widespread in pegmatite, especially in the Olary district of South Australia and in the Harts Range in the Northern Territory, but gem varieties are hard to find. Many pegmatites near Emmaville and Torrington in the New England district have yielded green (emerald) and blue (aquamarine) beryl crystals, while attractive aquamarine crystals occur with the topaz at Mount Surprise. Emerald crystals from localities in the Poona and Menzies regions of Western Australia occur in metamorphic rocks, typically mica schists. Some of the emerald deposits have been mined commercially for short periods.

◊ **5.19** *(top)* Topaz crystal (3.5 cm) from Stanthorpe, Queensland.

5.20 Faceted topaz (67 carats, 47 mm long) from Stanthorpe, Queensland.

MOUNT SURPRISE

Topaz, aquamarine and coloured varieties of quartz are found in alluvial gravels in creeks and gullies that are tributaries of O'Briens Creek, near Mount Surprise, about 200 km southwest of Cairns. The gem minerals have weathered out of pegmatite veins and cavities in the granite bedrock, the Carboniferous Elizabeth Creek Granite. The area has been extensively worked for alluvial tin. Topaz occurs as colourless, blue and yellow crystals and waterworn fragments in the wash, while aquamarine crystals up to 10 cm long have been found in cavities in the granite. The area has been designated for fossicking, with a licence required.

◊

(top to bottom)

5.21 Faceted topaz crystals from Mount Surprise, Queensland (9.35, 8.80, 11.70 carats) (largest is 16 x 9 mm).

5.22 Dravite crystal group (8 cm across) from Yinnietharra Station, Pilbara district, Western Australia.

5.23 Faceted elbaite (3.5 cm long) from the Giles pegmatite, Western Australia.

5.24 *(below)* Beryl (aquamarine) crystal (4 cm) from Mount Surprise, Queensland.

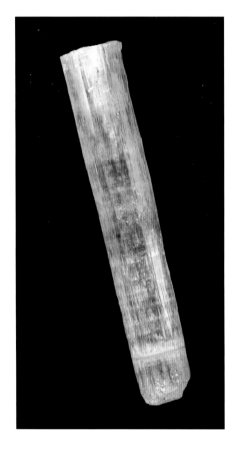

◊

5.25 *(top left)* Beryl (aquamarine)
crystal (4.5 cm) from New England,
New South Wales.

5.26 *(left)* Beryl (blue-green) crystal
(3.2 cm) from Torrington, New England,
New South Wales.

5.27 *(above)* Group of beryl (emerald)
crystals (largest is 18 mm long) from
Torrington, New England, New South Wales.

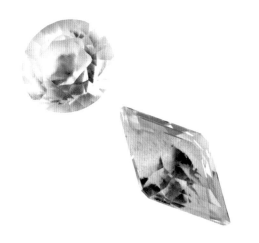

FLINDERS ISLAND

Flinders Island is the largest of the islands in Bass Strait between Victoria and Tasmania. Rugged granite outcrops dominate the scenery and in places show cavities that contain topaz crystals. Large topaz crystals that have been eroded from the granite into Killicrankie Bay, in the north of the island, have been called 'Killiecrankie diamonds'. Mostly waterworn by wave and current activity, the stones range from colourless and pale yellow to shades of blue, and make attractive gemstones when faceted (Bottrill *et al.*, 2004).

5.28 Faceted topaz (1.65 and 2 carats) (7 mm and 12 x 8 mm) from Killiecrankie, Flinders Island.

5.29 Waterworn topaz crystals (largest is 30 mm across) from Killiecrankie, Flinders Island.

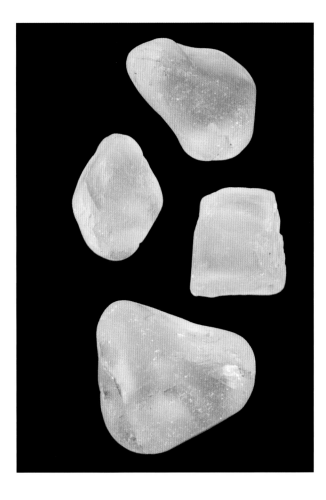

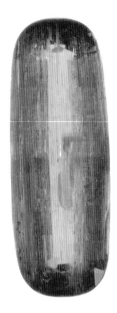

KANGAROO ISLAND

The Dudley pegmatite is situated near Penneshaw, on the eastern end of Kangaroo Island, and is associated with Cambrian granitic rocks. The pegmatite was mined for gem tourmaline (elbaite) during the early 1900s. Cavities and pockets in the pegmatite contained elbaite crystals displaying colours ranging from green to dark blue and pink, with some crystals showing the 'watermelon' combination of a pink core with a green rim (Keeling and Townsend, 1988; Webb, 2011). Pink and green gemstones have been faceted from these crystals.

5.30 Elbaite gemstone (16 mm long) from Kangaroo Island, South Australia.

5.31 Elbaite crystal portions (largest is 14 mm across) from Kangaroo Island, South Australia, showing 'watermelon' colour zoning.

POONA

Most of Australia's emeralds have come from the Poona area, in the Murchison district, some 600 km north of Perth (Schwarz, 1991; Bracewell, 1998). The Aga Khan mine has been the main producer but has been worked only intermittently. The emeralds are embedded in biotite schist, close to contacts with intruding pegmatite and quartz veins, and have formed due to contact metamorphism and chemical transfer during granite intrusion into Archaean metamorphic rocks. Hexagonal crystals up to 6 cm long have been collected, but most gems are cut from smaller crystals which have better colour and transparency. Emeralds in a similar geological setting come from the Menzies area, about 120 km north of Kalgoorlie.

◊

5.33 Beryl (emerald) crystals (up to 2 cm long) in mica schist from Poona, Western Australia.

◊

5.32 *(below)* Beryl gemstones, from left to right: aquamarine (3.5 carats, 7 mm), New England, New South Wales; emerald (0,75 carats, 6 mm), Poona, Western Australia; aquamarine cabochon (6.1 carats, 14 mm), Mt Surprise, Queensland.

Garnets

"Garnets were exhibited in very considerable quantities…among them several of good size and fine colour."

REV. JOHN BLEASDALE, 1867

◇◇◇◇◇◇◇◇◇◇◇◇◇◇◇◇◇◇◇◇

Minerals belonging to the garnet family are very widespread in many different igneous and metamorphic rocks throughout the world. Often they occur as well-formed crystals in shades of red, brown and green, depending on their chemical composition. Occasionally crystals may be free of flaws and have colours attractive enough to allow faceting as semi-precious gemstones. In Victoria, the garnet species almandine, spessartine, pyrope, grossular and andradite have been found; however, few crystals have been large or transparent enough to facet. This chapter describes only the main localities for each species; in some alluvial deposits several species from unrelated sources occur together.

Early history of discovery

As with many other Victorian gem minerals, garnets were first reported from the Ovens District goldfields around Beechworth in the 1860s. Here they occurred as fragments of red almandine in the washdirt from many of the creeks. George Ulrich (1867) noted their splendid deep claret-red colour, and suggested they would no doubt be valuable as gems if they were larger. A variety of garnets from the Ovens District was displayed at the Royal Society's exhibition in 1865. For example, George Stephen featured orange–yellow 'essonites' (probably spessartine), as well as almandine crystals and blood-red 'pyropes' (probably dark red almandine). Many garnets were also displayed at the Intercolonial Exhibition in Melbourne in 1866/7. Small garnet crystals found enclosed in smoky quartz pebbles from the Bradford Lead near Maldon

◊

6.1 Map showing main garnet locations in the text.

6.2 Almandine crystal (2.5 cm across) from Barnawatha.

were perfectly formed and a good colour, although as Ulrich noted, they were too small to be of value to the lapidary. The most notable discovery in the late 1860s was of large well-formed garnet crystals in granite forming the range now known as Mount Lady Franklin, near Chiltern, in northeastern Victoria. Later, during the 1870s, garnet crystals were reported at deep levels in the Eaglehawk mine at Maldon. Some Victorian garnets were faceted by local jewellers such as Spink and Son (Smyth, 1875), but surviving stones are extremely rare.

Numerous Victorian localities for garnets, particularly almandine, were listed by Atkinson (1897) and Walcott (1901). Many of these refer to isolated occurrences of garnets in granites or in alluvial deposits such as sands; they are not discussed in this Chapter.

6.3 Spessartine crystals and fragments from the Beechworth district (largest is 8 mm across).

Localities

NORTHEASTERN VICTORIA
Ovens District

Several species of garnet occur in the alluvial deposits of the Ovens District around Beechworth and Chiltern. Examination of concentrates collected last century by Edward Dunn from the creeks and deep leads in the district show most of the garnet grains are transparent shades of orange. As noted by Dunn in 1871, the full colour range is from dark red to pale orange–yellow, representing a variation in chemical composition from almandine to spessartine. Complete crystals are uncommon and most of the fragments are irregular and devoid of crystal faces. However some grains have retained striated faces characteristic of the trapezohedron, with edges which may be either sharp or abraded. A few crystal fragments show stepped dodecahedral faces. The maximum fragment size is about 8 mm, although larger pieces suitable for faceting have been collected.

The distribution of the garnets indicates that they have been derived from the local granite, especially from the tin lodes scattered throughout the district. Dunn (1871) recorded small brownish yellow trapezohedral crystals at Gimbletts tin lode, about 1 km northwest of Beechworth, and red crystals from Hensleys lode, about 6 km southwest of the town. On the basis of information provided by Dunn, Ulrich (1870) described the Gimbletts garnets as rounded grains and perfect crystals, from a pin's head to pea-sized, and a deep wine-yellow and semi-transparent.

A distinctive form of almandine garnet occurs as dark-red pitted crystals about 2 mm across. These are particularly abundant in concentrates from the Cocks Eldorado dredging operations near Eldorado, but also occur in some of Dunn's samples from other creeks in the district. The origin of these garnets is less clear and it is possible they are derived from the Permian glacial deposits near Wooragee.

There are no confirmed fragments of pyrope garnet, an indicator mineral for diamonds, in the alluvial concentrates from the Ovens District.

Barnawatha (Chiltern)

Garnets from this district were first noted by Smyth (1869a), who referred to a twinned crystal one inch in diameter from 'near Wahgunyah', sent to the Mining Department for inspection. A large imperfect, brownish red crystal, the size of a walnut, from 'the granite ranges near Chiltern' was described by George Ulrich in

6.4 *(opposite)* Almandine locality, Mount Lady Franklin, Barnawatha.

6.5 Almandine crystal (2 cm across) from Barnawatha.

1870. He also described dark greyish brown crystals found at 'Lady Franklyn Hill' and which had been presented to him by Edward Dunn. It is likely that all these occurrences are from the same locality — the summit of Mount Lady Franklin, 6 km east of the town of Barnawatha and about 12 km northeast of Chiltern. The peak has also been shown as The Kookaburra or Mount Kookaburra on older maps. The main rock type forming Mount Lady Franklin is a pale grey, fine-grained granodiorite containing small biotite flakes which show faint alignment (Leggo and Beavis, 1968). The granodiorite appears to have intruded the surrounding sedimentary rocks, which are now mica schists, during a period of metamorphism. Most of the large garnets occur in the granodiorite, with individual crystals widely spaced through the matrix. Smaller crystals are found in patches of granite pegmatite. The host rock is often weathered, resulting in a film of brown iron hydroxides on the garnet crystals, which may easily dislodge from their matrix.

The garnets are the largest and most attractive found in Victoria. Crystals may reach 5 cm across, although those exceeding about 25 mm in diameter tend to be distorted and show stepped faces. Crystals between about 5 mm and 2 cm across are more symmetrical, with the basic form being the trapezohedron. Modification more apparent as the diameter of the crystals increases. The colour ranges from brownish to purplish red, but as most of the crystals are heavily flawed, few are of gem quality. Analysis of a crushed crystal identified the garnet as almandine with a significant proportion of spessartine (Towsey, 1979). Further work on the compositional zoning and inclusions in these garnets is desirable to determine the conditions under which they crystallised.

Koetong

Fragments of pyrope garnet have been found in stream sediments from the Koetong area, where they are accompanied by cassiterite, zircon and ilmenite (Towsey, 1979). The crystals are sub-angular to rounded and up to 1 cm across. They are red to crimson and transparent to translucent. A small amount of chromium is present in the pyrope, which has a composition typical of garnets from ultrabasic rocks such as peridotite or kimberlite. This suggests the presence of a small volcanic pipe in the region, possibly intruding the Koetong granite of the region.

EAST-CENTRAL VICTORIA
Toombullup region

Two species of garnets occur with other gem minerals such as sapphire, zircon and topaz in the alluvial deposits in the Toombullup region (see Chapters 2, 3 and 5). However, only one, almandine, can be traced to local source rocks. These are Devonian volcanic rocks, mainly rhyolite and rhyodacite, which form a thick sequence in the Tolmie Highlands to the north of Toombullup. Almandine garnets are widespread in the volcanic rocks as dark-red crystals up to about 1 cm across. One theory on the origin of many of the garnets suggests they crystallised in metamorphosed sedimentary rocks buried deep in the Earth's crust. Melting of the metamorphic rocks provided the magma for the volcanic rocks, and the garnets were brought to the surface during the eruptions. Weathering releases the garnets into the local stream sediments,

where they are found as irregular fragments up to about 3 mm across. They are much less abraded than the accompanying reddish brown zircons, and often have pitted surfaces.

More enigmatic are beautiful, highly polished, transparent pinkish orange to plum-coloured pyrope grains up to 9 mm across. These appear to be concentrated in the Middle Creek deposits, but may also occur in other streams. Most of the fragments contain rutile inclusions. The pyrope grains are an important indicator mineral for diamonds but their source in the region is unknown. They appear similar to the pyrope from the Koetong district.

Dookie region

A thick sequence of mafic igneous rocks of Cambrian age forms the Ascot and Dookie hills, including Mount Major, in northern Victoria, about 180 km NNE of Melbourne. For many millions of years during the Paleozoic era, the rocks were buried by large-scale earth movements, and the resulting metamorphism formed dark-green minerals such as actinolite, chlorite and epidote. In places within these greenstones, as these rocks are called, hot solutions moved through fractures and reacted with the surrounding rocks. Resulting from this hydrothermal alteration are minerals containing boron, such as axinite–(Fe), datolite and danburite, as well as the calcium-rich garnet species andradite (Morvell, 1976; Birch, 1996). It is likely that the sample of ′green

6.6 *(opposite)* Faceted pyrope (4.5 mm long) from Toombullup district.

6.7 Waterworn pyrope crystals from the Toombullup district (up to 9 mm).

6.8 *(top)* Andradite crystals (up to 3 mm) from Dookie.

6.9 Andradite crystals (up to 2 mm) from Dookie.

garnet from the vicinity of Benalla', described by Smyth in 1874, was andradite from this region. There are several locations in the Ascot and Dookie hills where andradite occurs, but the best crystals have been found in the Shire quarry on Hooper Road, about 1 km northwest of the town of Dookie. At this locality, which is a designated Mineralogical Reserve, veins between 1 and 5 cm wide cutting through the greenstone contain crusts of pale yellowish green to honey-brown and dark-brown andradite crystals up to 5 mm across. The crystals are mainly rounded or distorted combinations of trapezohedral and dodecahedral forms, and are often intergrown. Accompanying minerals include axinite–(Fe), datolite, prehnite, actinolite and calcite. Dissolving the calcite in hydrochloric acid often exposes well-formed, frosted to lustrous andradite crystals lining the veins. Although many of the crystals are transparent, they are too small and flawed to be faceted.

Healesville–Eildon region

The mountainous region of the Cerberean Ranges, between Healesville and Eildon, consists of a thick pile of Devonian volcanic rocks, similar to those in the Tolmie Highlands in the Toombullup region. Garnets in the Cerberean volcanic rocks were first described from the Narbethong area by Junner in 1914. They are now known to be widespread in the rhyolite and rhyodacite units throughout the Cerberean Ranges, as dark pinkish red almandine crystals up to about 1 cm in diameter, but mainly less than 5 mm. These have a similar origin to the Tolmie Highlands garnets. The largest and best-formed crystals can be found in some of the granitic dykes which have intruded the volcanic rocks. However the garnets are too flawed for lapidary purposes.

WEST-CENTRAL VICTORIA
Bradford Lead

John Hornsby, a local mining identity and mineral collector, is credited with discovering small garnet crystals embedded in smoky-quartz pebbles in the Bradford Lead, near Maldon (see Chapter 5). During the 1860s, the lead yielded a range of gem minerals derived from the local granite (Ulrich, 1868). The garnets were initially considered to be rubellite (Bleasdale, 1868b), but when Hornsby sent some to the Geological Survey for examination, Ulrich recognised them as garnets. He described them as pin-head to pea-size, of a dark honey-yellow to dark blood and brownish red. The most symmetrical crystals were dodecahedrons, but many were modified by small faces of the trapezohedron. Ulrich described some unusual crystals, thick hexagonal plates showing concentric striations, which he explained by means of sketches. Specimens showing the range of garnet habits have been preserved in Museum Victoria's collections. Some were collected by Edward Dunn, who was Ulrich's field assistant during the Geological Survey's mapping of the area in the 1860s. Analysis of the garnets shows they have compositions between spessartine and almandine, but with slightly more manganese than iron.

Maldon

The Maldon area was rushed late in 1853 following the discovery of gold in gullies at the foot of Mount Tarrengower, about 20 km west of Castlemaine. Exploitation of the rich quartz reefs in the vicinity began several years later, and the Maldon goldfield was to become one of the State's richest, with 65 000 kg of gold won from the reefs. The quartz veins were in folded Ordovician sedimentary rocks metamorphosed to hornfels within the contact zone of the

FIG. 12 *a.*

FIG. 12 *b.*

◊

6.10 Garnet crystal sketches, Bradford Lead (Ulrich, 1870).

6.11 *(below)* Almandine crystal (5 mm) in quartz, Bradford Lead.

Harcourt Granodiorite. The veins were concentrated along several major north–south trending reef systems, with the easternmost being the Eaglehawk Reef, running through Union Hill at the northeast corner of the town. The earliest mines on this reef included the Eaglehawk United Company, which reached depths of over 1100 ft during the 1880s. At the 600-ft level, a large cavity was encountered. The only published description of this feature is by Moon (1897), who referred to it as a vugh lined with crystals of a variety of minerals, including garnet and amphiboles. The few specimens preserved contain calc-silicate minerals such as wollastonite, grossular, diopside and vesuvianite, which crystallised during the contact metamorphism. It is likely that the vugh was caused by dissolution of a mass of calcite associated with the calc-silicate minerals.

Only a few garnet specimens have survived, mainly through the efforts of Hornsby, whose collection of Maldon minerals was donated to the Museum of Victoria in 1948. They include dark red crystals up to 1 cm across associated with actinolite, diopside and calcite, and isolated dark reddish brown crystals to 1 cm embedded in quartz. These are trapezohedrons modified by dodecahedral faces. Crusts of transparent, pale brown dodecahedral crystals to 3 mm also occur. The only available analysis indicates the Eaglehawk Reef garnets are grossular.

Meredith

A weathered volcanic pipe containing pieces of hornblende, mica and garnet occurs 5 km east of Meredith, about 50 km west of Melbourne. The pipe, on private property just west of the Moorabool River, was first described by Ferguson in 1909. The surface expression of the pipe is a circular depression about 80 metres across in a black clay soil containing the mineral fragments. Drill cores from deeper levels show the rock in the pipe to be a volcanic breccia having some resemblance to kimberlite, but unlikely to contain diamonds (Day *et al.*, 1979). Of note are the two types of pyrope garnet, one pink, the other orange, occurring in the rock. Both varieties form small rounded to fractured grains up to 4 mm across. The orange pyrope also occurs as altered xenoliths up to 2 cm across. The pink pyrope contains more magnesium and chromium than the orange variety. While both varieties are transparent, the grains are too small to allow faceting as gems.

6.12 Grossular crystals (up to 3 mm) from Eaglehawk Reef, Maldon.

OTHER LOCALITIES

Small pink, mauve and orange garnets are abundant in streams draining Permian glacial deposits near Tylden. Many are well-preserved crystals with complex combinations of faces. Similar garnets derived from Permian sediments occur in the deep lead at Morrisons, where they are accompanied by orange and claret-coloured garnet fragments which may have come from volcanic pipes similar to that at Meredith. Claret-coloured garnet fragments occur in creeks near Daylesford, notably Jim Crow Creek and tributaries of the Loddon River. These garnets may be derived from xenoliths in the Fern Hill basalt northeast of Daylesford. In the beach sands and gravels of the Lake Bullenmerri maar complex near Camperdown, orange to claret garnet fragments are abundant. They have originated from garnet-pyroxene xenoliths which occur in the nearby volcanic ash beds.

6.13 Pyrope crystal fragments (up to 3 mm) from Meredith.

Garnets in other Australian States

Garnets of many compositions are widespread in metamorphic and granitic rocks in Australia. However, most crystals are highly fractured, so are not widely used as gemstones. There are many localities in the Harts Range where garnets crystals, usually deep red to pinkish almandine, can be found weathering from metamorphic rocks. In the early days these were thought to be rubies. Small gemstones of a few carats can be cut from fragments free of fractures. More recently, pale to deep orange crystals of grossular (in the gem variety known as hessonite) enclosed in quartz from near Coggan Bore in the eastern Harts Range have been faceted into attractive gems. Beautiful crystals of deep red spessartine, the manganese-dominant species of garnet, are a feature in coarser parts of the sulfide ore from Broken Hill, in New South Wales. Despite their smooth lustrous faces, the garnets are generally too dark and fractured and have too many inclusions to be used as gems.

◇

6.14 *(below)* Grossular crystal (2.5 cm) and almandine crystals (to 4.5 cm) from Harts Range, Northern Territory.

6.15 *(above)* Faceted almandine gemstone (10 mm) from Harts Range, Northern Territory.

◊

6.16 Spessartine crystals from Broken Hill, New South Wales (5 cm group).

6.17 (*above*) Faceted grossular gemstone (10 mm) from Coggan Bore, Harts Range, Northern Territory.

6.18 (*left*) Grossular crystals (main is 6 mm) from Coggan Bore, Harts Range, Northern Territory.

Quartz

"Smoky quartz and cairngorms, as they were called, were in great abundance, as also pure white crystals, and waterworn lumps, and amethyst of every shade of purple."

REV. JOHN BLEASDALE, 1865

✧✧✧✧✧✧✧✧✧✧✧✧✧✧✧✧✧✧

Quartz is the most common gem mineral found in Victoria, with the varieties rock crystal, amethyst, smoky and citrine all recorded. Because it is so common, it would not be feasible to describe all the quartz localities, so details are only provided for sites which are historically significant or have been popular with collectors.

Quartz with gem properties occurs in three main geological environments: quartz veins and reefs, often gold-bearing; in cavities in granites, where it may be accompanied by other gem minerals such as topaz and tourmaline; and in similar cavities, known as geodes or 'thunder eggs', in volcanic rocks. Quartz from reefs is generally milky to colourless, while that from granitic and volcanic rocks tends to be smoky or amethystine. Quartz eroded from all these rocks is often found as rounded crystals and pebbles in alluvial deposits. The main localities for quartz are discussed under these geological environments. Microcrystalline varieties of quartz, such as agate, are described in Chapter 8.

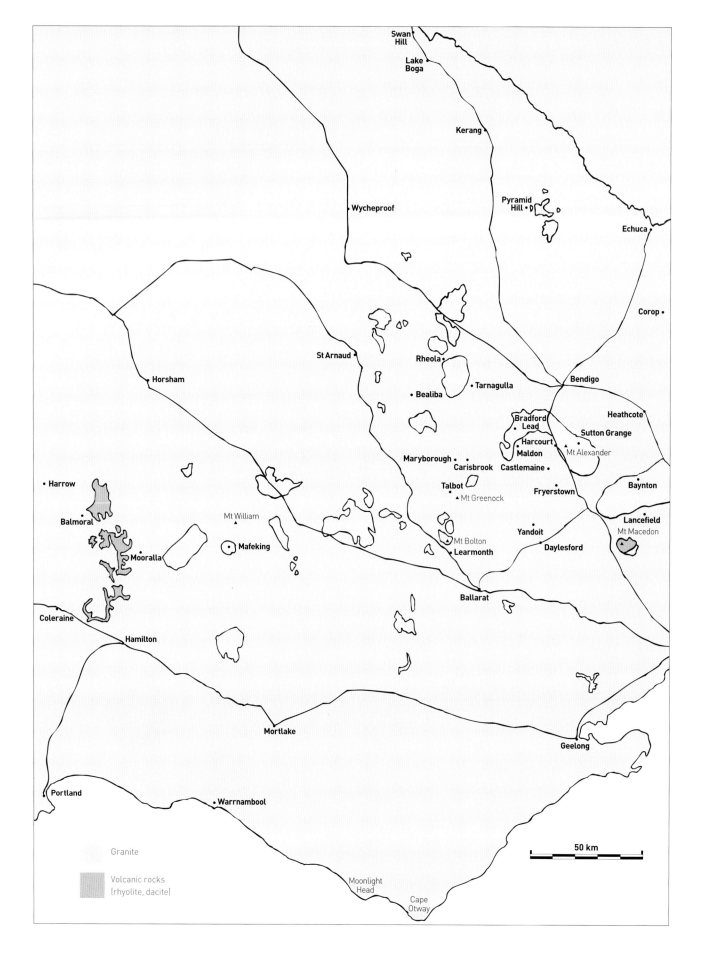

Swan
Hill
Lake
Boga
Kerang
Echuca
Wycheproof
Pyramid
Hill
Corop •
St Arnaud
Rheola
Bendigo
Heathcote
Horsham
Bealiba
Tarnagulla
Bradford
Lead
Sutton Grange
Harrow
Harcourt
Maldon
Mt Alexander
Maryborough
Carisbrook
Castlemaine
Talbot
Fryerstown
Baynton
▲ Mt Greenock
Balmoral
Lancefield
Mt William
Yandoit
Mt Macedon
Mafeking
Mooralla
▲ Mt Bolton
Daylesford
Learmonth
Coleraine
Hamilton
Ballarat
Mortlake
Geelong
Portland
Warrnambool

Granite

Volcanic rocks
(rhyolite, dacite)

Moonlight
Head

Cape
Otway

50 km

Early history of discovery

The discovery of quartz varieties, like many other Victorian gem minerals, is linked to the opening up of the major alluvial goldfields in the 1850s and 1860s. The conspicuous nature of quartz crystals made them readily collectable by the miners. Referring to the many varieties, the Rev. John Bleasdale noted at the 1866 Intercolonial Exhibition, *"these stones occur all over the Colony, but have been mostly found where digging for gold is carried on. Among the exhibits of Mr Turner, of Beechworth, were a number of exceeding beauty and size, exquisitely cut, and tastefully set as brooches and ladies' ornaments."* In 1874, Spink and Son, the Melbourne jewellers, cut 200 Victorian quartz specimens (Smyth, 1875), an indication of the popularity of the stones.

The earliest records of quartz crystals date from the Colonial Exhibition catalogues. The 1854 Melbourne Exhibition catalogue records smoky quartz from Black Range, Upper Goulburn River, and 'hexahedral quartz' from Perrys Run, south of McIvor. Ulrich (1866) sketched crystals of 'hood quartz' from veins in granite near Pigeon Hill, Tarrengower (now Maldon), and various forms from Fryerstown and near Baynton (see page 154). Crystals from the latter locality show well-developed 'phantoms'.

7.1 *(opposite)* Map of western Victoria showing localities described in chapters 7 and 8 (see Fig 8.1 for eastern Victoria).

7.2 Quartz crystal sketches (from Ulrich, 1866).

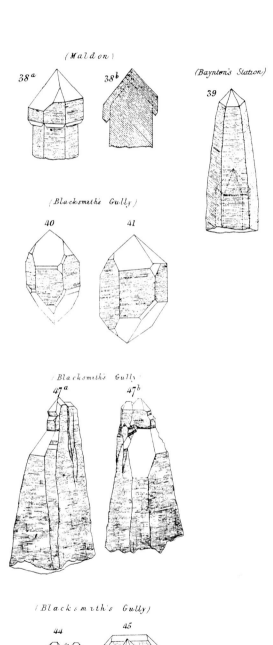

Quartz crystals from reefs and veins

Victoria has produced over 2 450 000 kg of gold, making it one of the world's richest gold provinces. As the rich shallow alluvial deposits and deep leads were exhausted in the 1860s, miners turned their attention to the ultimate sources, gold-bearing quartz reefs. Ulrich (1866) noted that *"druses of crystals occur in all the reefs, yet not, as one would have expected, in such abundance and beauty as in those ore-lodes of European mining countries, where quartz, as it does here, forms the lode-stone."* Ulrich also observed that pebbles of rock crystal were frequently found in the gold drifts on most of the goldfields, in particular Beechworth, Tarrengower (Maldon), Guildford, Avoca and Bendigo.

Typically, quartz derived from the reefs is milky due to the presence of fluid inclusions; clear crystals are uncommon. Abundant rounded pebbles of this milky quartz, derived by erosion of the quartz reefs, made up much of the alluvial deposits. If these deposits were on or close to granite bedrock, then amethyst, citrine, rock crystal and smoky quartz supplied by the granite could also be found in the gravels.

Despite the existence of thousands of quartz reefs across Victoria, very few fine quartz crystal specimens have been preserved.

Bendigo

The famous Bendigo goldfield was Victoria's richest, producing about 680 000 kg of gold between 1851 and 1954, mainly from quartz reefs after some early production from shallow alluvial workings. The reefs, which are hosted by strongly folded Ordovician slates and sandstones, are of many different types. Quartz crystals were collected from the various mines but, unfortunately, accurate location details were rarely documented.

Quartz crystals from the Bendigo reefs are stout lustrous prisms, typically milky white to colourless. They occur in groups associated with well-formed crystals of carbonate minerals such as dolomite, ankerite or calcite, occasionally with a scattering of pyrite. Albite and fluorapatite also occur with the quartz. The maximum size reached is unrecorded, but individual crystals on museum specimens reach 14 cm.

◊

7.5 Quartz crystals (to 27 cm) from Fryerstown.

◊

7.3 *(opposite, left)* Quartz crystal group (2.8 cm) from Blacksmiths Gully reef, Fryerstown, sketched by George Ulrich in 1866 (see #44 in Figure 7.2).

7.4 *(opposite, right)* Smoky quartz crystal (5.5 cm) with face polished to reveal 'phantom'; from near Baynton (see #39 in Figure 7.2).

Fryerstown

The gold-bearing quartz reefs near Fryerstown, on the Castlemaine–Chewton goldfield, produced some of the best quartz crystals found in Victoria. They were also amongst the earliest to be illustrated, with sketches by George Ulrich showing a range of unusual forms from the Blacksmiths Gully reef at Fryerstown (Ulrich, 1866). An array of quartz crystals from mines such as Rowe Brothers, Burdett-Coutts Company, Ferrons Reef, Duke of Cornwall, Clarkes Reef, Herons Reef and Blacksmiths Gully Reef at Fryerstown was exhibited at the 1886 Colonial and Indian Exhibition in London.

Fryerstown quartz crystals are milky white to clear prisms, often in randomly interlocking clusters. One large doubly terminated crystal preserved in the Museum of Victoria is 27 cm long. Another attractive specimen is a quartz crystal with inclusions of green chlorite encased in a transparent pyramidal termination.

Old mine dumps in the Castlemaine–Chewton region have been popular localities for collectors seeking quartz crystals.

7.6 *(below)* Engine house ruins, Duke of Cornwall mine, Fryerstown.

Woods Point area

Gold was discovered in the Woods Point region in 1860. Despite its remoteness, the goldfield became one of the State's most productive, with a few mines, such as the A1 at Gaffneys Creek, operating until the present day. The gold occurs in quartz reefs in numerous dykes, consisting mainly of diorite, and Middle Devonian in age (the Woods Point Dyke Swarm).

Ulrich (1866) noted fine crystals of quartz in the reefs at Woods Point and Raspberry Creek. In the Royal Society's Exhibition of 1865, J. B. Were exhibited quartz cut and mounted in a gold pin from the Age of Progress Mine at Woods Point. Dunn (1908b) described the A1 mine and noted that *"frequently spaces or 'vughs' occur in the reef and in these beautiful masses of quartz crystals are found"*. Attractive clusters of colourless quartz crystals up to 10 cm long, from the Morning Star mine in Woods Point, have been preserved in Museum Victoria. Small gold and pyrite crystals are found nestled amongst the quartz on some specimens. Fine specimens have also been collected in the Little Comet mine.

In the 1960s and 1970s, attractive groups of colourless quartz prisms could be collected in reddish soil in a track cutting near Matlock. The crystals were probably being released from a weathered diorite dyke.

Tarnagulla

Clear quartz crystals containing gold inclusions and sprays of rutile needles were recorded from the Crystal Reef, Stony Creek, Tarnagulla, by Smyth (1869b). The crystals are up to 12 cm long and may be doubly terminated. Two smoky-grey specimens of quartz showing 'Japan twinning' were collected from the Tarnagulla district in the early 1980s; the crystals are up to 5 cm across.

Durdidwarrah

Excellent crystals of colourless quartz were collected from a locality known as Durdidwarrah, near Meredith, by gem-club fossickers during the 1960s and 1970s. The crystals occurred in white clay-filled cavities in a large mass of reef quartz. Reaching 12 cm long, and commonly with unusual intermediate faces, the crystals are highly lustrous and transparent.

◊

7.7 *(opposite)* Quartz crystal group (3.5 cm across) from Little Comet mine, Woods Point.

7.8 *(top)* Quartz crystals (6 cm long) from Durdidwarrah (Collection of M. and L. Legg).

7.9 *(right)* Quartz crystals (to 2.5 cm long) with 'phantoms' of red and green chlorite from Rushworth State Forest.

Rushworth

Water-clear quartz crystals up to 5 cm long, enclosing unusual 'ghosts' or 'phantoms' of green and reddish chlorite, were discovered in the Rushworth State Forest, near Whroo, in 2012. They occur in cavities in thin quartz veins cutting the weathered basement sedimentary rocks.

Maindample

Attractive clusters of transparent quartz crystals up to 8 cm long have been collected recently from veins within Devonian sedimentary rocks (Norton Gully Sandstone) near Maindample. Many crystals have a slightly tabular habit and some contain greenish chlorite.

Quartz crystals from granite

Granitic rocks which contain cavities and coarse-grained patches of pegmatite are ideal environments in which to find quartz crystals. Suitable granites are widespread across Victoria. Most crystallised during the Silurian and Devonian periods, between about 400 and 370 million years ago.

7.10 Faceted amethyst from Beechworth (*top*, 1.5 cm across, and *centre*) and Starling Gap, near Powelltown (*bottom*, collection of Tony Forsyth).

NORTHEASTERN VICTORIA
Beechworth District

The greatest variety of gem quartz occurs in the Beechworth district. Pebbles and crystals of amethyst, citrine, rock crystal and smoky quartz (cairngorm and morion), as well as many unusual combinations, have all been found in gravels in the local streams forming the Eldorado Lead system. Many of the gemstones have been derived from the granite that forms much of the hilly country in the Mount Pilot Range north of the town (see Chapters 2 and 3 for more details on the history and geological features of the region). During the early gold and tin mining, many of the quartz varieties were found, not just in the gravels, but also in the granite bedrock of the streams. In more recent times, gemstone fossickers have been lured to Black Sand Creek and the Woolshed Valley down to Eldorado by the prospect of finding quartz crystals, as well as agate, topaz, sapphires and even diamonds.

Quartz is found along the length of the Woolshed Valley as far west as Eldorado. According to Dunn (1913), Black Sand Creek, which drains the granite hills to the north, was the main source of quartz pebbles (except for amethyst). Crystal groups weighing up to 10 or 15 kg were common at Sebastopol when mining

7.11 (*left*) Cabochons of amethyst (6 mm) from the Eldorado district.

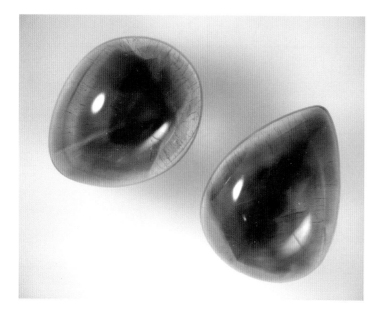

commenced in that area. Dunn also recorded occurrences of large quartz crystals within the granite at Reids Creek and Pennyweight Flat.

Perhaps the most sought-after quartz variety is amethyst. Dunn noted that it was not found in Black Sand Creek, but was abundant near Eldorado. Amethyst washed from clay veins in decomposed granite from the Old Wellington Company, Eldorado, was exhibited at the Colonial and Indian Exhibition in London in 1886. At Fergusons Gap, Dunn recorded prisms of grey quartz with amethyst hoods on the points. He was also the first to describe the beautiful amethyst crystals found in decomposed granite two miles (3.2 km) north of Reids Property at Eldorado. This locality is now known as Specimen Hill farm, near Springhurst. It was a popular fossicking locality over the years (Adams and Krantz, 1983) but collecting is no longer permitted on the property. Most of the amethyst crystals from Specimen Hill are doubly terminated, with very short prism faces, and are between 3 mm and 1 cm long. The colour ranges from pale lilac to dark purple, often with the colour concentrated in the pyramidal terminations. The crystals are mostly found loose in the soil, but some have been found attached to a matrix of weathered, fine-grained granite.

The most spectacular amethyst crystals, including sceptres, have

◊

7.12 *(above, right)* Amethyst crystals (up to 8 mm across) from Springhurst.

7.13 *(above)* Sceptre crystal of amethyst (7 cm) from the Eldorado district.

7.14 *(right)* Sceptre crystals of amethyst (4 cm) from the Eldorado district.

◊

7.15 Amethyst sceptre crystal on brownish quartz crystal (6.5 cm) from Eldorado.

7.16 Quartz crystals from Beechworth (*right*) and Bradford Lead (*left*, 26 cm, and *centre*).

been found in the granite in the Reedy Creek Track area, north of Eldorado. Magnificent sceptre crystals attached to plates of granite were obtained during several excavations in a quartz vein in the 1990s and early 2000s. The crystals were up to 10 cm long. Prismatic smoky quartz crystals up to 15 cm long were also found with the amethyst (Sutton and Roberts, 2011). Pyramidal crystals up to 3 cm long and an intense dark purple have been collected at Shannon Hill, Byawatha. Sceptre crystals consisting of colourless quartz coated with a thin layer of milky quartz and topped with an amethyst cap were collected from Pennyweight Flat, near Beechworth.

Citrine, referred to as 'false topaz' by Dunn (1871), is not as abundant as amethyst in the region. It is restricted to the alluvial deposits and has not been found as crystals in granite. It occurs as transparent, rounded, waterworn pebbles and broken fragments up to 6 cm across, with the colour ranging from pale lemon to golden sherry. The citrine was popular as a cut stone, with several pieces displayed by Rev. Bleasdale and the Melbourne jewellers Spink and Son at the 1865 Royal Society Exhibition. Pebbles of transparent light amber quartz from Deidrich's Sluicing Claim between Woolshed and Eldorado were exhibited at the 1886 Colonial and Indian Exhibition in London. Although scarce, citrine pebbles can still be found (Poth, 1984b). Specimens in the Museum Victoria collection have been faceted as beautiful oval and round cut stones. Smoky quartz, also called cairngorm or morion, was abundant amongst the gemstones discovered during early mining. At Reids Creek, a cavity in the granite yielded crystals up to 20 cm long, with black tourmaline (Dunn, 1871). At Sebastopol and Eldorado, smoky quartz crystals up to 9 cm long and coated with milky white quartz were collected. Smoky quartz crystals with pyramidal terminations coated by pale yellow–brown clay, and subsequently overgrown by colourless quartz, have been collected near Wooragee.

Yackandandah
Very large rounded crystals of amethyst were recorded from Rowdy Flat, at Yackandandah, 15 km northeast of Beechworth (Dunn, 1871).

◊

7.17 *(above)* Rough and faceted citrine and smoky quartz (up to 5 cm) from the Beechworth district.

7.18 *(left)* Intricately faceted citrine (522 carats, 5.3 cm across) from the Beechworth district.

◊

7.19 Large smoky quartz crystal (22 cm) from Terip Terip in the Strathbogie Ranges.

EAST-CENTRAL VICTORIA
Strathbogie Ranges, Crystal King mine

There are many localities for quartz crystals in the Strathbogie Ranges. Most of these are in stream gravels derived from the granite forming the ranges (see Chapter 5), but in some places fine smoky quartz crystals have been found in weathered granite. Several of the more productive localities include Creightons Creek and Hughes Creek near Ruffy, where terminated crystals over 12 cm long have been collected. Plates of fine smoky quartz crystals are reported to have been found in granite near Ruffy in 1996. Better documented is the discovery made on a property a few kilometres south of Terip Terip in 1998. A veteran fossicker found a clay-filled pipe-like deposit which he opened up by means of a shaft, and hauled out smoky quartz crystals up to nearly a metre across. Most of the larger crystals had been naturally damaged during past earth movements, but some fine crystals up to 40 or more cm long were recovered. A similar 'pipe' has recently been excavated in granite north of Bonnie Doon, yielding smoky quartz crystals and groups up to 50 cm across, commonly with a coating of muscovite flakes. Smoky quartz crystals also occur at the schorl localities near Longwood and Violet Town.

However, by far the best known quartz crystal location in the Strathbogie Ranges is the Crystal King mine. The mine is situated near Tallangallook, on the northern slopes of the ranges, at the head of Black Charlies Creek, about 14 km northeast of Bonnie Doon. Quartz crystals were first detected on the dumps of an old gold mine, known as Black Charlies. The mine was reopened in 1944 to search for quartz crystals suitable for piezoelectric applications such as frequency controls in radio transmitters, power stations and electric clocks and for use in microphones and gramophones. A second quartz crystal deposit was found nearby and the mine was renamed the Crystal King.

At the deposit, quartz forms cylindrical bodies within pipe-like masses of pegmatite which have intruded coarse-grained granite. Large crystals occur in cavities, or vughs, within the massive quartz pipes. The pipes and the surrounding granite are deeply weathered and the cavities are often full of clay.

Owen (1945) provided the earliest record of the quartz-mining operations, with further descriptions by Crohn (1952). The quartz was mined by a syndicate of five, including Mr John Willey, who prepared the piezoelectric plates in the Melbourne suburb of Caulfield. Willey's backyard later became a hunting ground for lapidary enthusiasts.

The workings consisted of five shafts sunk on pipes, and three trenches. Quartz crystals were also recovered from Black Charlies Creek. Shaft No.1 was sunk to approximately 30 m in a pipe which was roughly circular with a diameter of about 2.4 m. At a depth of 14 m, the pipe was intersected by a fault. Owen (1945) reported that a vugh above the fault yielded two tonnes of crystals, including three individuals weighing 350, 250 and 200 lbs (160, 114 and 90 kg respectively). A further tonne of crystals was recovered below the fault. Shaft No. 2 was sunk to a depth of about 15 m and produced approximately two tonnes of crystals from 80 tonnes of rock. The pipe was circular to elliptical, with a diameter between 1.8 and 3 m. Three quartz crystal bodies were located within this pipe. The depth and production of the other shafts are not known. During the 1980s and at times since, unsuccessful attempts were made by other companies or individuals to open the mine to obtain some remaining large quartz crystal specimens. It is now very difficult to find any quartz crystals on the surface as the location has been picked clean by fossickers (Anon, 1966; Cutter, 1967).

Typically, the quartz crystals were embedded in white to cream and pale brown clay which filled the vughs. The walls of the cavities were lined with massive, cloudy grey quartz. Owen reported that only one doubly terminated crystal was found (it was purchased from John Willey by the Museum of Victoria in 1982), suggesting that the crystals stopped growing once broken off the walls. The crystals range from colourless to smoky grey through to brown and near-black. They are usually milky at the base, becoming more transparent towards the pyramidal terminations. Several growth stages are evident in some crystals with successive growth zones becoming more transparent. Bubbles and fluid inclusions are common, as are inclusions of other minerals. There are also examples of misshapen, flattened crystals that resemble broken fragments, but close examination reveals that all the surfaces are made up of a mosaic of crystallographic faces. In rare cases, crystal faces may appear curved.

In 1971, Mr Alex Amess faceted a large Crystal King mine quartz crystal to produce the largest hand-faceted stone in the world at that time. Mr Amess recalled: *"For two years I searched for a quartz crystal big enough and clear enough from which to facet a very large stone. I contacted Mr John Willey, the man who had dug the Crystal Mine at Tallangallook in Victoria. I went to his home and searched through his heaps of crystal, and found one piece weighing 32 lbs and measuring 10 inches base to peak and 8 inches across flats. I immediately asked if he would sell the stone, but he insisted I first take it home and polish a couple of faces to see if it was any good. This I did, and found it to be clear except for a fracture right down the middle.*

7.20 Quartz crystal (15 cm) with unusual form, including curved upper face, from the Crystal King mine.

I then had to cut out the largest clear section. When I eventually cut it, I finished with a block of near-perfect crystal measuring 7 3/4 x 4 3/4 x 3 1/8 inches. The 3 1/8 inch being the depth of the stone, meant I had to cut either an oblong or oval cut. What finally settled the type of cut was the discovery of a couple of small inclusions in the corners. By cutting an oval stone, these could be removed.

Then I cut the stone, removing more than 1400 carats in the process. As was my practice, all angles were set by eye — a ticklish job with a large oval stone. Many was the time I cursed the fracture in the crystal that had forced me to cut an oval instead of a round stone!

Three and a half months later after 200 hours cutting and polishing, the job was nearly done. Maybe a few scratches still remained, but what's the odds? I set out to cut the world's largest faceted stone — not the most perfect!" (Amess, 1973).

The finished stone weighs 8510 carats (1.7 kg), measures 19 x 11 x 6 cm and has 196 facets.

◊

7.21 Alex Amess faceting the 'Crystal King'.

7.22 *(below)* The 'Crystal King' cut by Alex Amess (19 cm long).

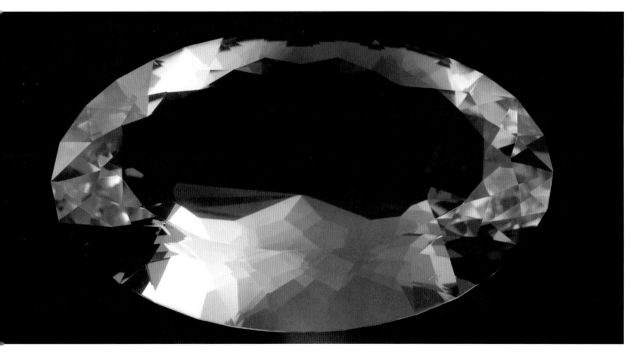

Black Range

Smoky quartz crystals up to 30 cm long were collected from an unknown location in the Black Range, north of Healesville, in the 1850s (see Chapter 5).

Lancefield district

Attractive gemstones have been faceted from pebbles of rock crystal and smoky quartz found in the headwaters of Mollisons Creek, northwest of Lancefield. These are probably derived from pegmatite veins within the local granite, the Cobaw Granite.

7.23 Selection of faceted rock crystal and smoky quartz from Mollisons Creek, near Lancefield, (largest is 3 cm across).

GIPPSLAND

Powelltown region

Several varieties of quartz crystals can be found in creek gravels in the Powelltown and Beenak region. The country rock is the Tynong Granite, which has yielded smoky quartz, rock crystal and amethyst, in crystals which may be of sufficient quality to facet. In places, smoky quartz crystals can be found in coarse patches within the granite, for example north of Starling Gap. Smoky quartz crystals up to 7 cm long and coated with whitish clay have been collected from the head of McMahons Creek. There are many other localities in the region east of Powelltown where quartz varieties can be collected, including streams near Noojee and Toorongo. In places the quartz crystals are accompanied by topaz and schorl crystals, also derived from the granite (see Chapter 5).

Phillip Island

The Woolamai Granite forms the spectacular and rugged pink cliffs of Cape Woolamai on the southeastern end of Phillip Island. The granite, of Devonian age, is one of the few pink granites in Victoria. It contains patches of pegmatite enclosing cavities lined with crystals of smoky quartz up to 8 cm long, as well as pink or red orthoclase and white to cream albite. A small quarry was operated on the eastern side of the cape during the early 1890s and most specimens were collected at that time.

WEST-CENTRAL VICTORIA

Baynton

While he was mapping the 5 NW Quarter Sheet in 1861, Norman Taylor, the Government Geologist, discovered attractive quartz specimens in granite at Jews Harp Creek, on Bayntons Station. The crystals were slender prisms up to 6 cm long and showed phantoms of milky quartz enclosed in smoky quartz (Ulrich, 1866, 1870). They were found lining the walls of a vein filled with yellow clay. Like the tourmaline occurrence described in Chapter 5, the locality is within the Cobaw Granite.

Bradford Lead

Large crystals and waterworn fragments of smoky quartz and amethyst, derived from nearby granite, were first collected in the alluvial gravels of the Bradford Lead, north of Maldon, during mining operations in the 1860s (see Chapter 5). Ulrich (1866), describing the amethyst, noted *"the light colour of specimens from this locality being often of such a character as to make the designation Rose Quartz more applicable."* Inclusions of garnet occur in some of the small smoky quartz crystals (see Chapter 6). In the late 1960s and early 1970s, the Bradford Lead was still a popular site with collectors, with quartz and topaz crystals being found. Spencer (1970) described a large smoky quartz crystal measuring 30 cm in diameter and 40 cm high, collected when a road was being bulldozed. Now the mine dumps have been thoroughly turned over and there is little indication of quartz crystals.

Mount Greenock, Talbot

Pebbles of amethystine quartz from wash dirt of the deep lead at Mount Greenock were displayed at the Colonial and Indian Exhibition in London in 1886.

Harcourt, Sutton Grange

Smoky quartz crystals were first recorded from a granite quarry on Mount Alexander, near Harcourt, by Hall (1895). The Harcourt Granite has been quarried for building stone up to the present day, but quartz crystals have rarely been encountered. However, quartz crystals weathered from the granite have been collected from Myrtle Creek, Sutton Grange, by lapidary clubs since the 1970s. Gemstones faceted from the crystals range from colourless to yellowish through to smoky brown.

Mount Bolton

Gravel pits at Mount Bolton, near Learmonth, have yielded smoky quartz and amethyst crystals weathering out of coarse pegmatite patches in the granite. Groups of amethyst crystals up to 14 cm high were collected in the early 1990s. Typically, the specimens consist of dull amethyst hoods overgrowing pale grey to smoky quartz. Smoky quartz crystals are more abundant and range from about 1 cm up to 12 cm long. They are commonly found with orthoclase, and rarely with beryl (see Chapter 5).

Rheola

Alluvial and reef mining operations on the Rheola goldfield from the 1850s unearthed several varieties of quartz. The Rheola Granite contained quartz veins and also 'nests of coarse crystals' (Dunn, 1890). Alluvium derived from the granite was a rich source of waterworn quartz crystals. Dunn (1890) reported that the Cenozoic gravels capping the Humbug Hills and Possum Hill contained considerable numbers of quartz crystals, rounded at the edges, cairngorms, some of unusual size and fine quality, citrine and amethyst. In addition, topaz and small garnets were recovered from the alluvials (see Chapter 5).

Bealiba

Deep leads in the Bealiba area was first mined for gold in the early 1850s. A large smoky quartz crystal showing growth zoning and with numerous inclusions of arsenopyrite was collected from No. 9 Hill at Bealiba, also referred to as Cochranes diggings. These diggings were opened up in 1858 (Flett, 1970). The crystal, now in Museum Victoria, was first described by Smyth (1869b). It was collected from a depth of 50 feet (16 m) from the surface and was lying in the auriferous deposit on the bed rock, which consisted of clay, slate and sandstone intersected by numerous small quartz veins.

7.24 Faceted smoky quartz (up to 4 cm) from Sutton Grange (collection of Edith Oakes).

NORTH-CENTRAL VICTORIA
Lake Boga and Pyramid Hill

In the granite quarry 10 km SSW of Lake Boga, cavities containing large crystals of quartz, feldspar, muscovite and fluorapatite, as well as other minerals, have been found (see Chapter 5). Quartz crystals are pale pinkish brown to dark smoky, translucent to transparent, and often of sufficient quality to be faceted. Lustrous, sharply terminated prisms up to 23 cm long have been collected, although most are between 5 and 8 cm long. At Pyramid Hill, quarrying of granite has also exposed pegmatite patches and cavities containing crystals of smoky quartz, orthoclase, albite and fluorapatite.

Topaz and tourmaline have been found in both quarries (see Chapter 5).

WESTERN VICTORIA
Mount William – Mafeking Goldfield

This goldfield, situated on the southeast edge of the Grampians Range, was discovered in 1897 and was subsequently rushed in 1900 (Flett, 1970). The gold was derived from quartz veins in the Mafeking Granite. Cenozoic gravels and decomposed quartz veins in the underlying granite were worked for gold (Murray, 1914), and amethyst crystals were found in both environments. Along with other quartz crystals, they were abundant in the alluvium and often showed a zonal structure. The miners believed that the distribution of the gold was associated with the amethyst. Hart (1905) reported that the quartz in the granite was in very thin veins for the most part, but swellings occurred lined with crystals of a somewhat amethystine colour.

7.25 *(above)* Faceted smoky quartz (2.5 cm) from Lake Boga.

7.26 Smoky quartz crystals (up to 4 cm) from the Pyramid Hill granite quarry.

Edward Dunn collected a 4-cm, doubly terminated crystal from a gravel capping on the granite at Nields Gully. The crystal, now in Museum Victoria, is uneven in colour and shows pale lilac veils. Other museum specimens from this locality include waterworn crystal fragments up to 10 cm long, usually with amethystine cores and colourless to milky rims.

Coleraine area

Deep-purple amethyst crystals up to 1 cm long were recovered from fine-grained granite in Konong Wootong Creek near Coleraine. Amethyst occurred in veins in sandstone in the Wartook Reservoir in the Grampians (Ferguson, 1917). Pegmatite veins in metamorphic rocks are shedding pieces of massive, fractured smoky quartz into the bed of Frenchmans Creek, near Balmoral. Smoky quartz pebbles, along with australites, found on the bank of the Glenelg River 5 km east of Harrow, are believed to have been carried there by Aboriginal people (Baker, 1958). At Nareen, massive smoky quartz has been collected.

Quartz in volcanic rocks

Rhyolite is the quartz-rich volcanic rock equivalent in chemistry and mineralogy to granite. There are many occurrences of rhyolite in Victoria, mostly formed as a result of violent eruptions from large caldera volcanoes during the Silurian and Devonian periods. In some places, hot fluids trapped in the cooling flows created cavities in which quartz was deposited, either as bands of agate, or as well-formed crystals. These cavities are often spherical and are referred to as geodes or, to use a more popular name, 'thunder eggs'. Geodes are more resistant to weathering than the surrounding rhyolite and may be preserved in the soil or on the surface.

WESTERN VICTORIA
Mooralla

Victoria's most prolific smoky quartz locality is situated near Andersons Gully, approximately 9 km northwest of the town of Mooralla. A designated gem-collecting site, the locality

7.27 Smoky quartz crystals (5 cm group; 6 cm geode) found at Mooralla.

is marked on the Beear 1:25 000 map and is reached by rough bush tracks. Such has been its popularity with fossickers for over 30 years that the sparsely timbered site is pockmarked with low dumps and partially in-filled pits (Price, 1979).

Keen prospectors have dug pits over five metres deep in weathered rhyolite, known as the Rocklands Rhyolite, which was erupted in the Late Devonian period. They are searching for layers of geodes containing the smoky quartz crystals. The vertical profile in a pit being excavated by a collector in 1996 consisted of a cream surface soil overlying a grey to red clay which contained white geode fragments. This persisted for 0.75 to 1 metre, before passing downwards into greenish grey weathered rhyolite with bands of geodes. The collector had dug down about 3–4 m and had been rewarded with several bucket-loads of geodes and crystal clusters. Most of the geodes from Mooralla contain a group of smoky quartz crystals which have become detached. While the crystals are very dark brown to nearly black, they are notable for their transparency and lustre. They often form complex intergrown groups up to 3 or 4 cm across and may contain large fluid inclusions. The most prized specimens are hollow spherical geodes up to about 10 cm across with a single crystal or a group of intergrown crystals attached to the inner wall of the geode. The walls may be lined with opaque white or faintly amethystine quartz prisms.

EASTERN VICTORIA

Quartz crystals have been found in agate-filled geodes weathered from rhyolite flows in the Snowy River area and also in the highlands north of Briagolong. The main localities are described in the next chapter.

7.28 Diggings for smoky quartz geodes, Mooralla (1996).

Quartz in other Australian states

Quartz in all its varieties, from colourless 'rock crystal' through citrine, smoky and amethyst, is widespread across Australia, and numerous localities provide fine specimens for both the gem cutters and crystal collectors. The White Rock quarry near Adelaide and the pegmatite deposits around Kingsgate in New South Wales are best known for transparent quartz crystals. Superb smoky quartz crystals come from the Torrington district of New South Wales, while three amethyst localities stand out — Wyloo Station in Western Australia, Soldiers Gap in Queensland and localities in the Wyalong Range, southwest of Katherine in the Northern Territory.

WHITE ROCK QUARRY

The White Rock quarry near Stonyfell, in the Mount Lofty Ranges east of Adelaide, has yielded magnificent colourless quartz crystals, as single crystals and clusters up to 60 cm across. The quarry produces quartzite for crushed rock, which is obtained from the thick Stonyfell Quartzite of Late Proterozoic age. The quartz crystals occur in cavities and veins in the quartzite and show many different habits, such as distorted growth, tapering, double terminations and twinning. They have formed during periods of hydrothermal activity accompanying metamorphism (Sutton, 2002).

7.29 Quartz crystal group (30 cm) from the White Rock quarry, Adelaide, South Australia (South Australian Museum collection).

◊

7.30 *(above)* Faceted quartz gems from the Kingsgate district, New South Wales. Largest stone is 52 x 40 mm; 566 carats.

7.31 Smoky quartz crystals (largest is 16 cm long) from the Torrington district, New South Wales.

WYLOO STATION

A deposit of amethyst crystals on Wyloo Station, about 150 km southeast of Onslow, has been a lure for collectors since the early 1900s, despite its remoteness (Bracewell, 1995). From its early days until the late 1960s, when it became known as the Great Australia Amethyst mine, the deposit was mined intermittently for high-quality amethyst for export. The amethyst crystals occupy the decomposed central portion of a dyke-like structure a few metres wide and over a kilometre long. The crystal groups are strongly coloured, but generally only the tips are of faceting grade. Another deposit of amethyst crystals occurs on Mt Phillip Station, about 150 km south of Wyloo.

◊

7.32 *(top)* Polished amethyst piece showing zoning (6.5 cm high) from Wyloo Station, Western Australia.

7.33 Amethyst crystal cluster (21 cm across) from the Wyalong Range, Northern Territory.

Chalcedony & Opal

"For chalcedonic varieties Beechworth is also the principal place of occurrences, as was exemplified by the fine collection of onyxes, carnelians and agates of diversified and beautiful patterns shown at the late Intercolonial Exhibition."

JAMES NEWBERY, 1868

◇◇◇◇◇◇◇◇◇◇◇◇◇◇◇◇◇◇◇

Quartz occurs in many varieties other than the well-formed
crystals described in Chapter 7. Most varieties consist of a hard
compact intergrowth of microscopic quartz crystals, known as
cryptocrystalline or micro-crystalline quartz (or silica), or more
generally as chalcedony. Sub-varieties of chalcedony, such as
agate, carnelian, onyx, bloodstone and chrysoprase, have been
named for their colours and patterns. While these names have
gained acceptance through common use in the art of jewellery
and lapidary for hundreds of years, they do not signify distinct
minerals, since they all consist of quartz. Recent research has
shown some very subtle differences in various properties between
quartz *per se* and some varieties of chalcedony. Another silica
mineral, moganite, is commonly present in small amounts in agate
and other microcrystalline forms of quartz, but it has not yet
been confirmed in any Victorian specimens.

Opal, or opaline silica, is a name commonly used for non or weakly crystalline forms of silicon dioxide which contain small amounts of water. Opal is considered to be a mineral distinct from quartz. Arrays of tiny spheres of silica make up the structure of opal; the relative sizes of the spheres and their crystallinity are used to classify opal into a number of varieties.

All cryptocrystalline quartz, as well as opal, formed by precipitation of silica from aqueous solutions in different geological environments. Of particular interest are pieces of silicified ('fossilised') wood that have been completely replaced by chalcedony or opal. Occurrences in Victoria are discussed according to the geological environments in which they have formed or are found.

8.1 *(opposite)* Map of eastern Victoria showing localities described in chapters 7 and 8.

Early history of discovery

Aboriginal people used chalcedony for thousands of years to make scrapers and spear tips. Fragments of various coloured chalcedony can often be found near campsites in coastal regions and on riverbanks throughout Victoria. The earliest written record in Victoria is of 'opal wood and opal raw' from Mount Macedon, displayed at the 1854 Melbourne Exhibition. However, it was during the 1860s, when alluvial gold mining was in full swing and geological mapping was underway, that a host of new discoveries were made. Angus McMillan made one of the first of these, whilst exploring the remote headwaters of the Moroka River in northern Gippsland in 1864. His journal records several localities where 'thunder eggs', or geodes consisting of agate, were found; ultimately some of the geodes were deposited in the collection of the Geological Survey. Many of the localities that were popular with collectors during the 1960s and 1970s were first recorded a century before. Agate and jasper from Cape Otway were exhibited at the Royal Society's 1865 exhibition. The shingle bed at the mouth of the Gellibrand River, west of Cape Otway, was known to contain agate (Ulrich, 1866) and there were other places along this rugged coastline which yielded polished pebbles of jasper and porphyry (Selwyn *et al.*, 1868). In 1866, George Ulrich noted that pebbles of chalcedony and carnelian were abundant in the creeks in the Beechworth district. He also recorded varieties of chalcedony from the Yarra Basin, Phillip Island, and the Murray River near Wodonga.

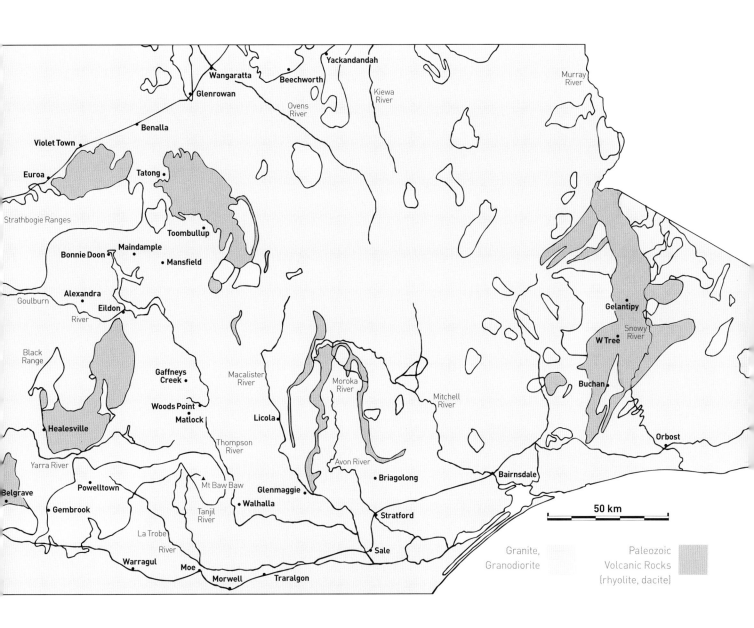

Yackandandah

Wangaratta

Beechworth

Glenrowan

Murray River

Kiewa River

Benalla

Ovens River

Violet Town

Euroa

Tatong

Strathbogie Ranges

Toombullup

Maindample

Bonnie Doon

Mansfield

Gelantipy

Alexandra

W Tree

Goulburn

Eildon

Snowy River

River

Black Range

Gaffneys Creek

Macalister River

Moroka River

Mitchell River

Buchan

Woods Point

Matlock

Licola

Healesville

Thompson River

Avon River

Orbost

Yarra River

Briagolong

Belgrave

Powelltown

Mt Baw Baw

Glenmaggie

Bairnsdale

Gembrook

Walhalla

Tanjil River

Stratford

La Trobe

River

Sale

50 km

Warragul

Moe

Granite,
Granodiorite

Paleozoic
Volcanic Rocks
(rhyolite, dacite)

Morwell

Traralgon

Fossil wood was commonly encountered during gold mining in deep leads during the 1860s, for example at Daylesford, Hepburn and Springdallah (Smyth, 1866b, 1869b). Ulrich (1866) reported silicified wood from the Barrabool Hills and the banks of the River Barwon near Geelong, at Sutherlands Creek near Maude, and from localities near Bacchus Marsh. Even localities near Buchan, in the far east of the State, were yielding specimens in the 1860s. Ulrich described opal from Gelantipy, and Smyth (1869b) recorded jasper from the Tara Range. Fossil wood from the Glenmore Ranges was exhibited at the Colonial and Indian Exhibition in London in 1886.

Some of these varieties of chalcedony were used by local jewellers of the day. For example, at the Royal Society's exhibition in 1865, William Turner exhibited 'petrification agates, mounted as brooches' from the Woolshed Valley. Couchman (1877) noted that 20 varieties of agate had been cut by George Spink of Melbourne.

Jasper, chert & flint

Water carrying dissolved silica commonly infiltrates porous rocks, replacing them with fine-grained quartz. Rocks such as jasper and chert, which in Victoria are commonly found as bands within altered Cambrian volcanic rocks, form in this way. Much of the chert consists of quartz which recrystallised from opaline silica derived from microscopic marine organisms known as radiolaria.

Flint is a light to dark-grey variety of chalcedony found as lumps and nodules in limestone. Flint pebbles are common along the Victorian coastline and may have been eroded out of Tertiary limestone. It is also possible that some were originally in ballast carried by sailing ships wrecked along the coast.

◊ **8.2** Red jasper from Corop (polished slab and sphere, 6 cm across) and Heathcote.

HEATHCOTE

Chert and jasper ranging in colour from bright red to yellowish brown, green and black can be collected in the Heathcote area. These ornamental stones are associated with the belt of Cambrian volcanic rocks, or greenstones, extending from Lancefield northwards through Heathcote. Popular prospecting sites have been around Sheoak Hill and near Ladys Pass, north of Heathcote, and in the hills east of Tooleen, where Dunn (1907) described a 'remarkable occurrence of beautiful red jasper', in places veined by white quartz.

COROP

About 50 km north of Heathcote, and near the same belt of Cambrian greenstones, is the town of Corop. Jasper was recorded from Lake Cooper, just south of Corop, by Atkinson (1897). The area was a popular fossicking site in the 1960s and 1970s for red to yellowish-brown jasper associated with the greenstones. Specimens take an attractive polish.

TATONG–HOWQUA

There is another discontinuous belt of Cambrian greenstones, known as the Mount Wellington Belt, in east-central Victoria. At the northernmost exposure of the belt, near Tatong, siliceous sedimentary rocks consisting of black and green vitreous chert and red jasper occur (Howitt, 1906a). Jasper cobbles collected from Blackbird Creek on the Howqua River, further south on the Mount Wellington Belt, were exhibited at the Colonial and Indian Exhibition of 1886.

◊

8.3 Chert cabochons from Morrisons (up to 4.5 cm) (Private collection).

MORRISONS

Pieces of chert showing colour bands of grey, orange and brown have been collected at Morrisons, a popular fossicking locality for sapphires, on the Moorabool River (see Chapter 3). The source of the chert, which has been fashioned into attractive cabochons, is uncertain. It may have formed by silicification of sediments beneath basalt, or could be derived from Permian glacial rocks nearby.

BUCHAN DISTRICT

An attractive variety of red jasper containing small white fossils occurs near the Basin, in the Buchan district. The rock has formed by replacement of Devonian limestone by silica.

Chalcedony & opal from granite

Whilst granite is abundant in Victoria, chalcedony and opal are rarely found in this environment. However, several localities have produced interesting varieties.

BEECHWORTH DISTRICT

The most unusual form of chalcedony associated with the granite near Beechworth (see Chapter 2) is in the form of enhydros or 'water stones'. First found in the 1850s, then rediscovered by Edward Dunn in 1864, the enhydros were enclosed in clay at the contact between granite and sandstone near Pennyweight Flat. They aroused considerable interest, with some being displayed at the Royal Society's exhibition in 1865, and with both Ulrich (1866) and Dunn (1870) providing descriptions. The enhydros are irregular polyhedra which vary in length from less than 1 cm up to about 12 cm. While they closely resemble crystals, their flat faces show no consistent symmetrical relationships. The enhydros are usually hollow, with the walls consisting of opaque dark brownish-yellow to transparent pale-yellow chalcedony, ranging in thickness from wafer-thin up to 6 mm. The interior surfaces may be smooth, botryoidal, or lined with small quartz crystals. Water, sometimes containing gas bubbles, is present in some of the enhydros and was first analysed by Foord (1871). Several theories have been put forward to explain how the enhydros formed. The most likely origin, suggested by Cooksey (1895) and Dunn (1913), is that they formed from thin layers of silica deposited on the walls of cavities between intergrown tabular calcite crystals. The calcite crystals were later dissolved, leaving the chalcedony infillings. It is not known when the enhydros formed. Nor is the source of the silica-rich solutions which deposited the chalcedony known, although Dunn (1870) suggested these may have moved through a nearby fault. Other ideas on the origin of the enhydros were discussed by Boutakoff and Whitehead (1952).

In other occurrences in the district, thin veins of pale-yellow to red chalcedony were found in the granite at Black Sand Creek, Reids Creek and elsewhere (Dunn, 1913). At Reids Creek, pink opaline silica occurred as irregular veins in the granite and was studded with broken pieces of black tourmaline.

◊

8.4 Chalcedony 'enhydros' from Spring Creek, Beechworth. Largest specimen is 8 cm long.

GLENROWAN

Granite forming the Warby Range is being quarried at Glenrowan, about 15 km southwest of Wangaratta. From time to time, thin seams of precious opal have been found on fractures in the granite. The opal shows a bright play of colours, mainly blue, green and purple, but is not of gem quality. This is the only confirmed locality for precious opal in the State, although a pebble of gold-bearing quartz containing seams of precious opal was found in gravel near Mansfield in the 1990s.

8.5 Precious opal seam in granite from Glenrowan (field of view is 7.5 x 5 mm across).

Agate & chalcedony in volcanic rocks

Some silica-rich volcanic rocks, in particular rhyolite, can be a source of agate, usually filling geodes or 'thunder eggs'. These result from the infilling of gas bubbles by silica deposited from hot fluids as the rhyolite cools. Fluctuations in the amount and chemical composition of the fluids often lead to successive bands of silica with different colours. If the agate does not completely fill the cavity, quartz crystals may occupy the remaining space. Agates are resistant to weathering and tend to accumulate in soil formed from the surrounding rhyolite. They can also survive transport by rivers over long distances. The main agate-bearing volcanic rocks in Victoria are the Wellington Rhyolite in east-central Victoria and the Snowy River Volcanics in far eastern Victoria.

Much less common is chalcedony found within silica-poor volcanic rocks such as basalt.

BRIAGOLONG AND MOROKA DISTRICTS

Some of the State's best localities for agates are in the remote country north of the town of Briagolong, in east-central Victoria. They occur within the Wellington Rhyolite, which outcrops along the margins of a 60-km-long basin-like structure known as the Avon Synclinorium. The rhyolite is contained within a thick sequence of sedimentary rocks, including conglomerates, laid down in rivers and lakes on flood-plains in Late Devonian to Early Carboniferous times. The rhyolite shows flow-banded, porphyritic and spherulitic textures (Marsden, 1988). Agates may be collected from rhyolite outcrops and also from the beds of streams which drain through the region.

George Ulrich was the first to describe agates from the Wellington Rhyolite, based on some fine geodes collected from the Moroka Valley, at the northern end of the basin, by Angus McMillan during his exploration of the region in 1864: *"Cut in halves they all present pentagonal shaped concentric alternating striae, partly irregular mixtures of pale-bluish and opaque white chalcedony, each of the angles of the pentagon having apparently served as channels of infiltration for siliceous water."* (Ulrich, 1866). Identical geodes to these were rediscovered at places along or near McMillan's original track during the early 2000s. Specimens collected from Doolans Plains, near Moroka, are rough spheres up to 10 cm across, containing translucent white, grey and reddish chalcedony.

There are several thunder-egg localities in the Wellington Rhyolite near Briagolong. Agates can be collected in bushland off Duffys Road, about 8 km northeast of the town. The geodes occur in weathered rhyolite and a pale-brown clayey soil. Typical geodes are spherical, 5 to 6 cm across, and filled with milky blue-grey agate with white banding. At Red Hill, spheroidal geodes contain reddish-brown agate, commonly enclosing star-shaped cavities lined with white quartz crystals. At Cooks Road, small thunder eggs filled with milky agate may be found.

◊ **8.6** Agate in geodes up to 9 cm across, from Duffys Road, north of Briagolong (Private collection).

8.7 *(left)* Exposure of agate-bearing geodes in rhyolite, McMillans track, Moroka district (Photograph courtesy of Les Hooker).

8.8 *(below, left)* Agate geode (9 cm across) from Red Hill near Briagolong.

8.9 Geode with agate and quartz (9 cm across) from Licola district.

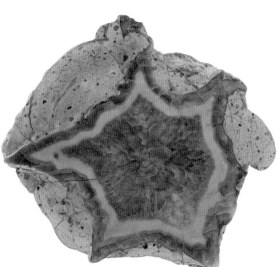

8.10 *(top)* Agate geode (13 cm across)
from Mitchell River, Tabberabbera.

8.11 Agate geodes (to 11 cm across)
from Mitchell River, Tabberabbera.

TABBERABBERA

Geodes up to 20 cm across
containing bright red and white
chalcedony have been found in
the bed of the Mitchell River at
Daisy Point, near Tabberabbera,
37 km northwest of Bairnsdale.
These have weathered from
one or more nearby horizons of
spherulitic rhyolite interbedded
with Late Devonian sediments
of the Avon River Group.

BUCHAN DISTRICT

The picturesque region centred on
the town of Buchan in far eastern
Victoria has been a popular
destination for gemstone fossickers.
Records of agates, jasper and
fossil wood go back to the 1860s,
but most collecting took place in
the 1960s and 1970s. Modern-day
visitors relied on the extensive local
knowledge of the late Gwen and
Mick Butterworth, who displayed a
fine collection of the region's rocks
and minerals at Stonehenge, their
caravan park at South Buchan. The
main collecting localities were
described in a brochure written by
the Butterworths and Leona Lavell
in 1969, but otherwise references to
minerals of interest to fossickers are
few (Dineen, 1969; Butterworth, 1971).

The geology of the region is complex,
but is dominated by the thick Snowy
River Volcanics, of Devonian age.
These were erupted into a north–
south-trending rift valley about
100 km long. The volcanic rocks are
well-exposed in the spectacular

gorges of the Snowy River and its tributaries, north of Buchan. Rhyolite is the main rock type and there are several localities where geodes containing agates can be collected. The main site is in rugged bushland and can only be reached by 4-wheel-drive vehicle, via the New Guinea Jeep Track. The 'thunder eggs' can reach up to 60 cm in diameter but are usually less than 15 cm across. They contain translucent agate in various shades of grey and reddish to yellowish brown, but without prominent banding. There is often an irregular central cavity lined with small quartz crystals. The best collecting in the region is done on the pebble beds in the Snowy River and its tributaries. For example, it is possible to find agate, carnelian, and 'ribbon stone' in the Murrindal River near the Pyramids mine and further downstream near the Basin Road crossing. Near the junction of the Buchan and

8.12 Agate geode (20 cm across) from the Snowy River, northeast of Buchan.

Snowy rivers, at a locality known as 'million yards of stone', and further upstream along the Snowy, agates in rhyolite can be collected. Most of the varieties of chalcedony are probably derived from the Snowy River Volcanics, but could have travelled many kilometres downstream.

MOORALLA
The geodes in the Rocklands Rhyolite near Mooralla, in far western Victoria, are best known for their beautiful smoky quartz crystals (see Chapter 7). Some of the geodes are filled by translucent pale reddish, brown and grey chalcedony.

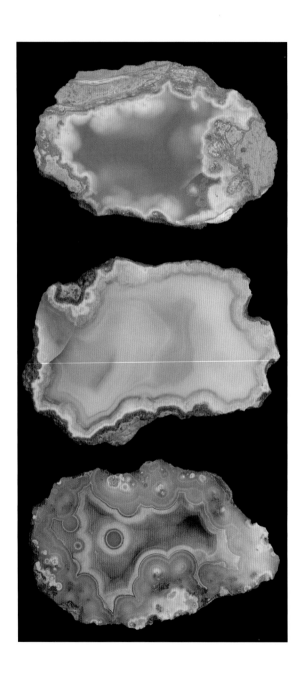

PHILLIP ISLAND

Chalcedony from Phillip Island was first recorded by George Ulrich in 1866 and numerous specimens were collected by the Geological Survey (Selwyn *et al.*, 1868). The chalcedony occurs as translucent white, cream or pale bluish-grey linings to irregular cavities in weathered basalt belonging to the Tertiary Older Volcanics. The most popular fossicking locality for chalcedony on Phillip Island has been Kitty Miller Bay, on the rugged south coast (Smith, 1967b). At the base of the cliffs on the eastern side of the bay, veins of translucent, milky-white to bluish-grey chalcedony occur on the basalt shore platform. These veins extend seaward below the water level, making collecting difficult and at times dangerous. There are also vughs lined with colourless quartz crystals or filled with translucent colourless to pale grey, faintly banded agate, some with linings of carbonate minerals. Rough waterworn pieces of the chalcedony can also be found on the shore platform. Some of the chalcedony is thick enough and of sufficient quality to be shaped into milky-looking cabochons.

Banded calcite–ankerite pebbles resembling eye agates also occur at this locality (see Chapter 9).

Silicified wood

Tertiary alluvial gravels and deep-lead deposits across the State commonly contain pieces of fossilised wood. In most samples, the wood has been replaced by silica deposited in the original pore-spaces from solutions percolating through the old buried stream beds. This process often resulted in the structure of the timber being preserved, so that features such as growth rings can still be seen. The type of silica varies from common opal and chalcedony to porous varieties. The silicified wood found in Victoria ranges from large sections of tree trunks to parts of branches. Victorian specimens have been included in studies of how wood is replaced by silica (Scurfield and Segnit, 1984).

The main localities where silicified wood can be found are described in the next two sections.

8.13 *(opposite)* Polished agates (up to 8 cm) formed in basalt from near Kitty Miller Bay, Phillip Island. The chalcedony also contains white to cream and pale brown carbonate bands.

Common opal

Common opal in Victoria has usually been formed when volcanic glass was altered by groundwater during cooling of lava flows. Most of the Victorian examples are associated with basalt flows erupted during the Tertiary period. There are many localities where common opal has been collected (Atkinson, 1897), but only the main ones are described in this section.

GELANTIPY DISTRICT

George Ulrich described the occurrence of opal in the Gelantipy district in east Gippsland. In 1870 he wrote: *"Another new locality for opal is the basalt of an outlier at Gelantipy, Gippsland. It is there abundantly dispersed through the rock in small rounded grains, some of which, by showing some play of colours, resemble the real precious opal very much."* Specimens were placed in museum collections (Selwyn *et al.*, 1868), and the deposit was being worked by two miners late in 1873 (Smyth, 1874). Ferguson (1902) reported on the occurrence and mentioned that other localities existed in the region.

The largest common opal deposit is on private property at W Tree, just off the Gelantipy–Buchan road, between Murrindal and Butchers Ridge. The opal occurs in vesicular Tertiary basalt resting unconformably on the Snowy River Volcanics (Teichert and Talent, 1958). The stone is flawed and jointed at the surface, but unweathered specimens are vitreous to compact with a conchoidal fracture. They may be beautifully streaked and mottled in yellow, brown, red, black and milky white. Waterworn pieces of wood replaced by common opal have been found in the vicinity.

SUNBURY–RIDDELLS CREEK DISTRICT

Selwyn (1861a) and Selwyn *et al.* (1868) recorded opal jasper occurring as a band of irregular lumps and nodules in basalt at Riddells Creek, 45 km NW of Melbourne. Ulrich (1870) reported that: *"Fine nodular pieces of [opal] have been found in the basalt near Sunbury. One specimen, in the Public Museum collection, is coloured light-blue in the centre, and greenish-brown outside, showing in this outer division parts that are white, vesicular, and friable, and contain iron pyrites."* Most of the common opal collected more recently in the district is compact and yellowish-brown. A locality in Emu Creek, approximately 3 km east of Sunbury, may be the one referred to by Ulrich. Atkinson (1897) recorded opal from Woodend and Gisborne but gave no details.

MORWELL DISTRICT

Atkinson (1897) listed opal from Moe and Morwell but without description. A basalt quarry on Creamery Road, approximately 5 km from Yinnar, near Morwell, has been a popular collecting site for brown to black and mottled common opal and opalised wood. Common opal has also been collected from a dam site at Driffield, about 7 km southwest of Morwell.

DURDIDWARRAH

Common opal was recorded underlying basalt in the Parish of Durdidwarrah, near Bacchus Marsh, by Selwyn *et al.* (1868), but there are no descriptions of the material.

CLUNES

In the 1960s and 1970s, dark-brown vitreous common opal was collected from under basalt in Tullaroop Creek, near Clunes.

8.14 Common opal from various localities: Riddells Creek *(left)*, Butchers Ridge *(rear)*, Moe district *(right)* and Yinnar (with cabochon). Largest piece is 11 cm across.

Chalcedony, agate and jasper pebbles in alluvial deposits

The beautiful rounded pebbles of agate, chalcedony, jasper and other varieties of cryptocrystalline quartz found on beaches and in river beds throughout Victoria have had complex histories, involving several cycles of erosion, transportation and deposition. Many of the pebbles are derived from local rocks, such as those described above, but there are also many which have originated much further afield. These foreign pebbles were carried to the region by glaciers during the Permian period and dropped when the glaciers melted. Once very widespread, the deposits left by the glaciers have been mostly removed by erosion. This resulted in most of the pebbles being redistributed into younger sedimentary rocks. However, they can still be found in a few small patches of Permian glacial sediments preserved in western, central and northeastern Victoria.

BEECHWORTH DISTRICT

Agate, jasper and other varieties of chalcedony are common in the Beechworth district. Many pebbles were unearthed during gold mining along Reedy Creek last century, but because they can still be found today, the region has long been popular for fossicking (Poth, 1984b). With few exceptions, the pebbles of agate, chalcedony and jasper found in the Reedy Creek gravels have been eroded out of Permian glacial sediments. The largest occurrence of these rocks is near Wooragee, close to the headwaters of Reedy Creek. Edward Dunn was the first to deduce the source of the pebbles, by noting their distribution along the creek from Wooragee down to Eldorado (Dunn, 1871, 1913). He also found rough agates in other patches of glacial conglomerate near Eldorado and Springhurst.

8.15 Agates up to 7.5 cm across from the Beechworth district.

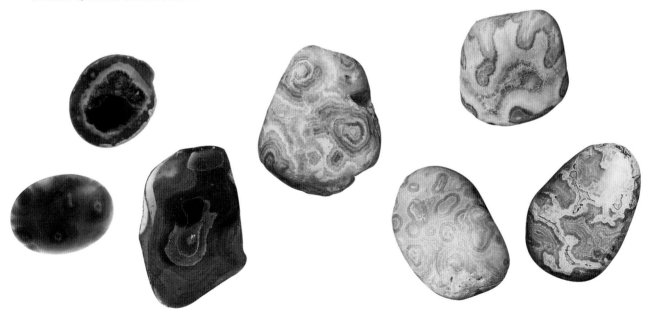

In the first descriptions of agates from the Beechworth district, George Ulrich noted that *"specimens of fine size, colour and pattern are very often found; banded varieties, the bands arranged in the peculiar zigzag shape; 'Fortification agate' as also 'Landscape' and 'Moss agate' being not rare amongst them"* (Ulrich, 1866). Most of the agates found along Reedy Creek have indistinct bands and rings showing subdued colours, mainly greyish to cream, yellow and brown. They are always very smooth and well-rounded. Those from Eldorado tend to be the smallest, having travelled further down the Woolshed Valley from the source rocks near Wooragee. Dunn (1871) reported that agates at Wooragee frequently weighed between 10 and 20 pounds (4.5–9 kg). A large agate boulder weighing 65 kg, with a circumference of 120 cm, was found 5 km from Eldorado in 1975 (Anon., 1975). The agate was donated to the Eldorado Museum.

Jasper pebbles, with colours of yellow, red, brown and black (the variety 'lydianite') are also found along Reedy Creek from Wooragee to Eldorado. Some of the bright red jasper pebbles are veined with white quartz.

An unusual report of opal from the Beechworth district was made by Bleasdale in 1865. He wrote: *"White and milky but with a fair share of fire — I have seen in Beechworth some fine specimens much waterworn and in shape resembling rather long and flat French beans. Fire opal, I have seen only one specimen, which was given me by a Beechworth digger. It is a very grand one."* Bleasdale displayed a fire opal mounted in a ring at the Royal Society's exhibition in 1865, describing it as *"perhaps the finest fire opal yet found in Victoria, which I obtained from a digger at Beechworth, and got cut and polished by Mr Spink. This particularly fine stone, one part of which was cut off for a specimen,* *attracted the curiosity of the curious in these matters, for its size and beauty."* No specimens of fire opal from Beechworth have been preserved, or reported since.

GLENROWAN DISTRICT
Other Permian glacial deposits occur in northeastern Victoria, between the settlements of Greta and Taminick, in the Glenrowan district. In the south, near Greta, Kitson (1903a) mapped exposures of glacial deposits between the King River and Fifteen Mile Creek. In the most northerly deposit, near Mundara, boulders of black jasper (lydianite), chert of various colours, grey quartzite and agate occur. Many of the pebbles are polished, and show glacial grooves and scratches.

At Taminick, about 6 km northwest of Glenrowan, the glacial deposits are in a valley lying between Futters Range on the east and the Mokoan Ranges on the west. The largest inlier forms Cannings Hill, where the deposit consists of about 15 m of clayey gravels with loose pebbles of agate, quartzite, chert, granite and schist. It rests on Ordovician sedimentary rocks and is partly overlain by a basalt capping. Nearby, to the north and west, similar deposits occur on Sadlers Hill and Coxs Hill.

BUCHAN–NOWA NOWA DISTRICT

Large pieces of fossil wood, including trunks up to several metres long and over 50 cm in diameter, have been found in Tertiary alluvial deposits in the Buchan district. The wood has been replaced by porous silica, which usually appears cream to dark brown. Conglomerates containing jasper are exposed in cuttings along the Gorge Road, 1.5 km northeast of Nowa Nowa. Agate has been found as rolled lumps in Boggy Creek, near Nowa Nowa (Clark, 1892).

MITTA MITTA

Jasper was collected from the Italian Point lead on the Mitta Mitta River near its junction with the Gibbo River.

AVON RIVER, STRATFORD AREA

The broad valley of the Avon River near Stratford has been a popular collecting area with fossickers. At Weirs Crossing, a few kilometres north of the Princes Highway, and further upstream near Ridleys Lane, the river flows through extensive bars and banks of coarse pebbles. These consist of a variety of colourful siliceous, volcanic and metamorphic rock types that are found in the highlands to the north. Amongst the pebbles are banded agates, occurring as smooth, usually oval stones ranging from a few centimetres up to 17 cm across. The agates are grey to cream with brown to black bands and eyes forming intricate patterns. Red and yellow jasper pebbles are also abundant in the river bed. The Avon River and

8.16 Pebble beds in the Avon River near Stratford (1996).

its tributaries, Valencia Creek and Freestone Creek, drain through the Wellington Rhyolite to the north. Although the Avon River agates do not resemble any of the geode-filling agates from localities near Briagolong and Moroka, the rhyolite is a likely source. Cambrian rocks in the Mount Wellington region may be the source of the jasper. It is possible that the pebbles have been through several cycles of transport and deposition, including conglomerates of Late Devonian–Carboniferous age.

Waterworn pebbles of banded agate similar to the Avon River stones occur in high-level gravels near Newry, 15 km WNW of Stratford. These may have been carried down from the highlands to the north by the Macalister River. Similar agates have also been collected from gravel quarries near Dutson, 25 km SSW of Stratford. Red jasper and agate pebbles occur in Tertiary gravels at Toms Creek, approximately 25 km southwest of Bairnsdale.

8.17 *(above)* Agates (up to 7 cm across) from Newry, near Stratford.

8.18 Polished agate (10 cm across) from the Avon River at Stratford.

SOUTH GIPPSLAND

The picturesque South Gippsland hills are composed of Early Cretaceous sandstones and mudstones, with minor conglomerates near the base of the sequence. The sediments were deposited by broad river systems in a large rift valley which opened up between Australia and the Antarctic during the late Jurassic period. The valley extended across southern Victoria, and similar sedimentary rocks form the Otway Ranges, between 150 and 200 km to the west. Other details on the geology of the South Gippsland region appear in Chapter 3.

Associated with the Cretaceous sediments are numerous localities for agate, jasper and chert pebbles. Kitson (1903b) found naturally polished pebbles in agglomerate surrounding eroded volcanic centres at Townsend Bluff, Andersons Inlet, near Cape Patterson. At nearby Savages Hill, he described similar pebbles of agate, chalcedony, carnelian, chert, jasper and silicified wood in conglomerate. Cuttings along the Toora to Foster road, between Chitt Creek and westwards to Deep Creek, for a distance of approximately 3 km, exposed a conglomerate layer containing polished and striated pebbles. These ranged from 1 cm up to 45 cm in diameter and consisted of quartzite, dense hard fine-grained contact rocks, quartz, jasper and chert of various colours including black ('lydianite'), and granite (Ferguson, 1906b). At Bindings Hill, near Jumbunna, pebbles of jasper were recorded. Pieces of red and green jasper were found during gold mining in Livingstone Creek, a tributary of Turtons Creek (Ferguson, 1936). Banded agates have also been found in the latter creek. Many of the pebbles within the conglomerates and younger stream gravels may have been transported and polished by glacial action during the Permian period. They were subsequently eroded, reworked by river systems and deposited as conglomerates early in the Cretaceous period.

In South Gippsland there are localities where chalcedony and silicified wood can be found in Tertiary gravels beneath the basalt flows and in young streams. At Scotts Creek, a tributary of the Foster River, west of Jumbunna, blocks of silicified wood with veins of carnelian and chalcedony have been collected (Kitson, 1917). During the 1970s, a quarry at Ruby, near Leongatha, was a popular fossicking spot for agate, petrified wood and chert. Lapidary club members also regularly visited the Berwick quarry, for fossil wood found in gravels preserved beneath basalt. Silicified wood could also be collected from the Lang Lang River, along with rare small sapphires (see Chapter 3).

DANDENONG RANGES–GEMBROOK

Some of the sapphire-bearing streams in this region (see Chapter 3) also carry small pieces of coloured silica minerals. George Ulrich first recorded these in 1866, when he referred to 'cats eye', onyx and sardonyx, jasper and carnelian from the Yarra Basin. He also noted in 1870 that William Wallace Creek was rich in nodules and pebbles of chalcedonic varieties of quartz. Smyth (1872b) listed fire opals, sapphires, garnets and other precious stones which had been collected from the head of the 'Woori Yaloak River'. Agate, carnelian and opal from the river were exhibited at the Vienna Exhibition of 1873. Atkinson (1897) listed agate from Beenak and Sassafras Creek. The original source of the agate and chalcedony is unknown.

During fossicking activities by lapidary clubs in the 1960s and 1970s, rough pieces of agate and carnelian could be collected in William Wallace, Cardinia, Menzies, Hardys and Selby creeks. The stones are up to 5 cm across and range from opaque dull cream and brown, to translucent grey, orange and reddish-brown. Banded varieties are quite common. The pieces look very attractive when polished.

LODDON VALLEY, CARISBROOK

Permian glacial deposits are known from a number of outcrops in the Loddon Valley and from the Loddon Deep Lead between Creswick and Newbridge. They appear to have been preserved by down-faulting. Mines on the Chalks Deep Lead, near Carisbrook, uncovered a variety of pebbles, some of them striated, in tillite (Macumber, 1978). Red and grey granite, gneiss, sandstone and chert

were noted. A few very large agates showing brown wand white banding have been found in the deep lead.

YANDOIT

Pieces of a chalcedony geode from Yandoit, showing concentric banding enclosed by crystalline quartz, were exhibited at the Colonial and Indian Exhibition in London in 1886. Several specimens have been preserved in Museum Victoria collections. The specimens may have come from Permian glacial deposits near Campbelltown. Harris and Thomas (1948) described glacial sediments containing pebbles and boulders of agate, red granite, gneiss with tourmaline crystals and rhyolite in the Campbelltown area, notably at Yandoit Hill, a few kilometres west of the township.

8.19 Polished agates (up to 5 cm) showing banding from Cardinia Creek.

8.20 *(bottom left)* Large agate (21 cm across) from Chalks No. 3 mine, Carisbrook.

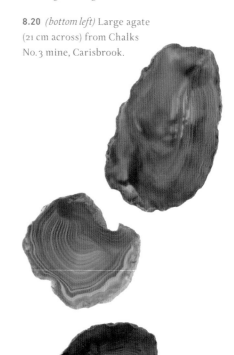

CAPE OTWAY, MOONLIGHT HEAD

The spectacular rugged coastline stretching from Cape Otway to Port Campbell has attracted tourists and gem-hunters over many years. The steep cliffs that form most of the coast are composed of a thick sequence of Cretaceous sandstones and mudstones, similar to those forming the ranges of South Gippsland. Tertiary marine and non-marine sediments overlie the Cretaceous rocks (Tickell *et al.*, 1992).

The favourite collecting localities have been beaches and bays near Moonlight Head, about 25 km west of Cape Otway. Wreck Beach, Moonlight Beach, Devils Kitchen, Dilwyn Bay and Pebble Point are the main sites and are reached via steep steps or paths down the cliffs. Care should be exercised with tides. All these localities are now within the Otway National Park, so that fossicking is no longer possible. A host of well-rounded pebbles can be seen in shingle beds at Devils Kitchen and Pebble Point. At other localities, the pebbles were obtained by sieving the beach sands (Anon., 1966a; Dineen, 1981). The most common pebbles are of porphyritic volcanic rocks such as rhyolite, none of which is exposed in the region. As well, several types of agate, jasper, silicified wood and many colourful siliceous pebbles have been collected. The most sought-after agates were christened 'moonies' by collectors. They are up to a few centimetres across, and consist of a core of translucent yellow, orange, red, white or black chalcedony with a white to creamy-brown skin. These stones are rough-tumbled or cabochoned until an attractive pattern of skin and coloured chalcedony emerges. Too much tumbling removes all the skin. Other agates

8.21 Pebble Beach, near Moonlight Head (1995).

show subtle banding in cream and brownish colours. Waterworn pieces of silicified wood, in shades of grey, cream and dark brown, also occur. A few exceptionally large pieces up to 40 cm long have been found. Amongst the other gemstones reported from the Moonlight Head coast are topaz, garnet, tourmaline and zircon (Stone, 1967). Waterworn fragments of green and brown glass may be from bottles washed from the numerous shipwrecks along the coast.

The source of the Otway-coast pebbles is uncertain, but they may have a similar origin to the pebbles found in the Cretaceous conglomerates in South Gippsland. They may have been eroded out of older conglomerates and carried to the coast by the Gellibrand River. As well as Cretaceous conglomerate beds in the Otway region, there are also pebble layers in the Tertiary sediments overlying them. For example, at a gravel quarry on

Wonga Road, about 12 km south of Colac, pebbles of grey and white banded chalcedony, chert and flint up to 10 cm across are abundant. These are possibly within the Wiridgil Gravel, which outcrops east of the Gellibrand River along the flank of the Otway Ranges (Tickell et al., 1992). The unconsolidated gravel is up to 70 m thick and contains minor pebble layers. The pebbles are of various rock types including Palaeozoic black shale, quartzite, rhyolite, granite and schist.

COLERAINE DISTRICT

There are reports of pebbles of jasper, chalcedony and silicified wood occurring in alluvial deposits in far western Victoria. However there are no published descriptions and few locality details are available. There are several potential sources for these gemstones in the region, in particular the Rocklands Rhyolite for agate and chalcedony, Cambrian greenstones with associated chert and jasper, and Permian glacial deposits containing a variety of pebbles. In the Coleraine district in particular, thick Permian sediments are preserved in the valleys of Koroite and Konong Wootong creeks. Further east, common opal and carnelian have been collected in a creek bed on the Glenelg Highway between Wickliffe and Glenthompson.

Chalcedony and opal in other Australian states

Opal is the dominant form of non-crystalline silica in Australia, with world-famous deposits of gem-quality precious opal at Andamooka, Coober Pedy and Mintabie in South Australia; Lightning Ridge and White Cliffs in New South Wales; and the Quilpie and Yowah areas of Queensland. There is also a myriad of varieties of colourful cryptocrystalline silica, or chalcedony, such as agate, jasper, chrysoprase, chert and silicified wood, with the legendary Agate Creek deposits in Queensland providing the most beautiful gemstones. Important deposits of bright-green chalcedony, coloured by nickel contents occur in the Marlborough district of Queensland and in the Kalgoorlie region of Western Australia. The best-known varieties of silicified or 'petrified' wood come from the Chinchilla district in Queensland; in the Kennedy Ranges inland from Carnarvon in Western Australia (the variety known as 'peanut' wood); and petrified 'man fern' from the Lune River region in southern Tasmania.

◊

8.22 *(opposite)* Polished agates (up to 7 cm across) and 'moonies' from Moonlight Head (Museum Victoria collection except for larger pieces at rear, in private collection).

8.23 Polished pebbles of petrified wood from Moonlight Head (to 7 cm) (Private collection).

AGATE CREEK

Beautifully coloured and patterned agates are found in the Agate Creek area, south of Georgetown in northern Queensland. The agates have weathered out of Carboniferous basalt at the head of three creeks — Blacksoil, Spring and Agate creeks. The agates are generally less than about 7 cm long and are oval shaped. Most display concentric banding on a fine scale, with prevalent colours of brownish-yellow, red and purple. Agates with quartz-filled or hollow centres are not as common as solid stones. Despite their great diversity, agates from particular localities in the fossicking areas may show diagnostic colours and patterns. The Agate Creek field is a designated fossicking area requiring a licence.

◊

8.24 Selection of polished agates (largest is 10 cm across) from Agate Creek, Queensland.

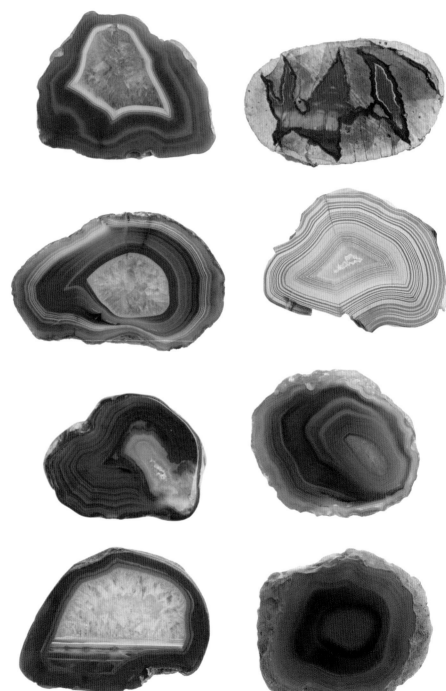

MARLBOROUGH

The finest chrysoprase in Australia has
been mined near Marlborough, about
100 km northwest of Rockhampton in
Queensland. The chrysoprase has formed
in veins and patches resulting from extreme
weathering of nickel-bearing serpentinites.
Colours range from very pale green to
translucent bright apple-green and opaque
bluish-green. The best quality chrysoprase
contains nearly 2.5 weight percent nickel.

8.25 *(below)* Chrysoprase in the rough
from the Kalgoorlie district (15 cm long).

8.26 Chrysoprase pieces (up to 4 cm)
from Marlborough, Queensland.

QUILPIE AND YOWAH DISTRICTS

The famous Queensland 'boulder' opal occurs in the weathered and silicified Cretaceous sedimentary rocks of the Winton Formation, between Cunnamulla and Winton in far southwestern Queensland. The opal fills random cracks and concentric fractures in siliceous 'ironstone' concretions, which range from 'nuts', only a few centimetres across, to boulder-sized. The opal tends to show blue and purple base colours, with flashes of red and green.

PRECIOUS OPAL FIELDS OF SOUTH AUSTRALIA AND NEW SOUTH WALES

Many varieties of precious opal, the gemstone synonymous with Australia, have been found in the famous fields of Coober Pedy, Andamooka and Mintabie in South Australia, and Lightning Ridge and White Cliffs in New South Wales. In all these fields, precious opal occurs as veins and patches, and occasionally as replacements of fossil bones and shells and of crystals, within Cretaceous sedimentary rocks associated with the Great Artesian Basin. The opal has formed during deep weathering of the sediments during the Tertiary period, when silica was released from the sediments and slowly redeposited, initially as a gel, in cracks and voids. Precious opal is classified and valued on such features as the predominant colour in the 'flash', the intensity and brilliance of the colour pattern, and the colour and transparency of the matrix. These give rise to terms such as light or 'white', dark or 'black', 'fire', 'jelly' and 'crystal'. Light opal is the most common, with Coober Pedy and Mintabie the main sources. High-quality black opal from Lightning Ridge is amongst the rarest and most valuable of gemstones.

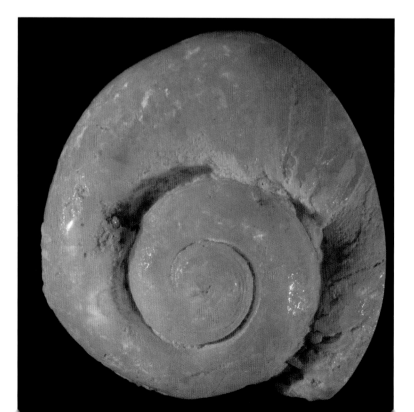

CHINCHILLA

Beautiful red and yellow silicified wood is found in abundance at localities near Chinchilla, 300 km west of Brisbane in the Darling Downs district. Fragments of various sizes and textures can be found in several sites on private land, whose owners permit licensed fossicking. The pieces come from gravels capping low ridges of Jurassic to Early Cretaceous sedimentary rocks. Chinchilla wood is used for a wide variety of decorative pieces.

8.27 *(opposite, top)* A spectacular specimen of 'boulder' opal, named 'The Galaxy', from the Queensland opal fields (10 x 8 cm). This piece is now in the collection of a museum in the USA.

8.28 *(opposite, below)* Precious opal replacing gastropod (1.5 cm across) from Coober Pedy, South Australia (Private Collection).

8.29 Precious opal replacing a cluster of ikaite crystals (9 cm high) from White Cliffs, New South Wales.

8.30 Polished slice of petrified wood from Chinchilla, Queensland, showing well-preserved growth rings (12 cm).

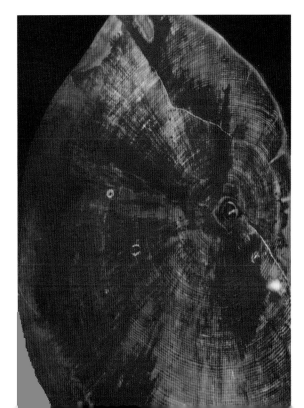

Other Gem Materials

"If the intimation from Victoria can be accepted as genuine, that the true turquoise has been found there in quantities little short of fabulous, that colony will begin a new era of prosperity."

Glasgow Evening Times, 30 OCTOBER 1893

◇◇◇◇◇◇◇◇◇◇◇◇◇◇◇◇◇◇◇◇◇

This chapter describes other Victorian minerals
which can be considered as gemstones or ornamental
materials. They are found in diverse geological
environments not covered in previous chapters.
Some of the minerals, such as turquoise, have been
collected to shape and polish, but others have merely
been reported without description.

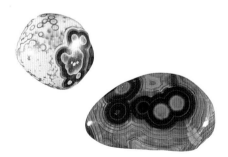

9.1 *(above)* Map showing locations in text.

9.2 *(opposite)* Polished cabochons of banded calcite-ankerite from Riddells Creek (brownish) and Kitty Miller Bay, Phillip Island (pale). Largest cabochon is 3.5 cm across.

CALCITE–ANKERITE

Attractively banded varieties of iron-bearing carbonate minerals occur in cavities in basalt at several localities in Victoria. These were first recorded in the 1860s, as 'ferro-calcite' (Selwyn *et al.*, 1868), from basalt flows at Barfold and near Keilor. Edward Dunn collected similar specimens a few kilometres north of Clunes. The best-known occurrence, however, is in a railway cutting in basalt just west of Riddells Creek, about 45 km northwest of Melbourne. At this locality, rounded cavities up to 5 cm across may be filled with compact carbonate minerals showing fine-scale concentric banding, closely resembling agate. Colours of the bands are cream, yellowish brown, orange-brown and dark brown. X-ray diffraction tests on the bands suggest most of them are made up of ankerite or ferroan calcite, but determination of their chemical composition is needed for more precise identification. Cabochons prepared from these concretions often reveal intricate patterns, with multiple 'eyes'.

Small calcite concretions up to 3 cm across can be found weathering out of basalt at Kitty Miller Bay, on the south coast of Phillip Island. These show concentric agate-like banding in shades of cream, grey and brown.

These banded carbonate concretions probably formed from solutions rich in carbon dioxide during the final stages of cooling of the basalt flows.

CASSITERITE

Despite its hardness (6.5 on the Mohs scale) and brilliant lustre, cassiterite is not generally considered to be a gem mineral. However, in many of the world's major tin deposits it often occurs in fine crystals, usually pale to dark brown to black, which may be partly transparent

and capable of being faceted. Cassiterite is very widespread in Victoria with many rich alluvial deposits and a few sub-economic lode deposits in granitic rocks (Cochrane, 1971). The orange to red variety, sometimes known as 'ruby tin', occurs in heavy mineral concentrates in several of the alluvial fields, notably the Ovens District and at Mount Hunter on Wilsons Promontory.

Dunn (1871) first described the varieties of cassiterite and their distribution in the creeks of the Ovens region. The grains are usually well rounded, up to 1 cm across, and range in colour from pale yellow through orange, red and dark brown to nearly black. Grains showing several colours are common. While the colours of transparent fragments are intense and attractive, there is usually too much internal fracturing for them to be faceted for gemstones. The presence of a few sharp or only slightly abraded crystals in some concentrates shows that the source of the cassiterite is the granitic rocks of the Beechworth region. Several small lodes were prospected in the region as early as the 1860s (Ulrich, 1870).

CORUNDUM

In some gemfields in Victoria, sapphires and rubies, the gem varieties of corundum, are accompanied by pebbles of opaque corundum. These are not derived from volcanic rocks, but are most likely to have come from corundum formed during metamorphism of other rocks. Some varieties may have been brought into the region by glaciers.

9.3 Pebbles of purple corundum ('barklyite') and blue dumortierite (6.5 cm across) from the Beechworth district.

Ovens District

The alluvial deposits in the Eldorado and Chiltern–Rutherglen deep lead systems (see Chapters 2 and 3) contain a wide variety of opaque corundum pebbles. According to Dunn (1871) these weighed up to 4 lb (1.8 kg) and came in shades of red, blue, brown and yellow, as well as black, grey and white. The most conspicuous variety formed smooth, often flattened pebbles, from a few millimetres to 7 or 8 cm across and ranging from bright pink to deep purplish-brown. Some of the small stones are translucent. This variety was recognised very early in the gold mining history of the region and was given the name 'barklyite' by George Stephen, in honour of Sir Henry Barkly, a former Governor of Victoria. The corundum pebbles are distributed from Yackandandah, down the Wooragee and Woolshed valleys as far as Eldorado (Dunn, 1871, 1913). Corundum pebbles collected by Dunn from the Chiltern Valley mine (Dunn, 1887), north of the local divide represented by the Pilot Range, include barklyite as well as grey and blue varieties. There are also small pieces of blue-grey corundum altering to muscovite, a feature characteristic of corundum from metamorphic rocks. Dunn (1913) concluded that the corundum pebbles were not of local origin, but were most likely to have been eroded out of the glacial conglomerates in the Wooragee area. However, this could not account for the corundum pebbles in the Chiltern Lead, unless these were recycled from Tertiary alluvial deposits which have since been removed by erosion.

Large blocks of pink to dark-purple corundum occur near the junction of the Wellington and Dolodrook rivers, near Licola in Gippsland, some 140 km south of Beechworth. The corundum is associated with serpentine-bearing rocks, or greenstones, of Cambrian age (Thiele, 1908; Dunn, 1909). Similar pinkish corundum-bearing blocks occur in Cambrian greenstones near Heathcote (Skeats, 1909). An analysis of the Wellington corundum obtained by Thiele and Dunn showed 1.4 weight % chromium oxide (Cr_2O_3), similar to contents in barklyite pebbles. This evidence suggests that the barklyite was derived by erosion and subsequent transportation of corundum associated with the Cambrian greenstones, possibly by glacial action.

Other localities

Small pebbles of fine-grained greyish corundum occur in the alluvial deposits of the Italian Hill deep lead near Daylesford (see Chapter 3). Small pieces of black corundum were said by Ulrich (1867) to occur in the drifts of many of the goldfields, including the Dandenong goldfield and at Ballarat (Selwyn et al., 1868). These were the variety 'emery', as confirmed by an analysis made by Newbery of the Dandenong mineral, which showed substantial iron was present, and by X-ray diffraction tests on the Ballarat mineral. The source of the emery is unknown. Corundum occurs as 2–4 mm, pale-pinkish hexagonal cleavage fragments with irregular bluish zones in concentrates from Sedimentary Holdings' alluvial gold operations at Amphitheatre, near Avoca. Similar mottled blue-grey corundum grains occur in the diamond and sapphire-bearing concentrates from Carapooee (see Chapters 2 and 3). These fragments originated from regionally metamorphosed rocks.

CHRYSOBERYL

Stone (1967) reported that chrysoberyl occurred rarely at Beechworth, but no descriptions have been published. There were early references to sapphires with chrysoberyl-like green colours, and predictions of chrysoberyl being found, but no verified specimens are known from Victoria.

DUMORTIERITE

The rare boro-silicate mineral dumortierite has been described from only a few occurrences of metamorphic rocks and pegmatites, only one of which is in Australia. It usually forms purplish or blue masses which are sufficiently hard to take a polish. Dumortierite was identified forming royal-blue aggregates in a waterworn pebble 7 cm across from the Wooragee Creek, in the Beechworth district. Edward Dunn probably found the pebble, and labelled it as 'sapphire-bearing'. Another smaller pebble, 1 cm across, with identical mineralogy, was found in one of Dunn's Beechworth gem-gravel samples in the Museum's collections. The dumortierite pebbles are clearly foreign to the region and it is likely they were transported to the Wooragee district by glacial action during the Permian. A glacial transport

◊

9.4 Rough and cabochoned 'selwynite' (round cabochon is 1.5 cm across) from the Heathcote district.

method is supported by the presence of rare grains of dumortierite in concentrations of heavy minerals from Permian glacial sediments at Bacchus Marsh (Jacobson and Scott, 1937). However, the ultimate source of the dumortierite-bearing rocks is unknown.

MOONSTONE

Moonstone is a gem variety of the feldspar mineral, orthoclase, characterised by an attractive silky-blue to grey schiller. This effect arises from interference caused to transmitted light by thin plates of another feldspar mineral, albite, interlayered with the orthoclase. Ulrich (1867) reported that small pieces of moonstone with weak chatoyant reflections occurred at Reids Creek, near

Beechworth. None of these have been preserved in collections.

'SELWYNITE'

A compact, dark-green, variously mottled rock type found near Heathcote was first described by George Ulrich in his 1866 Exhibition Essay. He believed it was a new mineral and named it 'selwynite' after Alfred Selwyn, the Director of the Geological Survey of Victoria. Ulrich tried to calculate a chemical formula for selwynite, based on incomplete analyses (no potassium was determined) by James Newbery. It was shown to be a rock, not a mineral, after more detailed chemical and mineralogical studies (Skeats, 1908; Watson, 1922; Mahony, 1922).

The main constituents of selwynite are diapore and muscovite, forming a very fine-grained mixture which also contains tiny, sparsely scattered chromite crystals. The proportions of these minerals vary, which explains the sometimes mottled appearance. Although the diaspore contains a little chromium, it is uncertain whether this mineral or the muscovite (possibly the chromium-bearing variety fuchsite) is responsible for the dark-green colour of the rock. White to cream veins cutting through the selwynite also consist of muscovite. Ulrich thought this was a new mineral as well and named it 'talcosite'. The rock is easily carved and takes a reasonable polish, but is difficult to shape into large ornaments because of closely spaced fractures. This has not dissuaded lapidary clubs from collecting the material over many years.

The original outcrop of selwynite was on the western slopes of Mount Ida, about 3 km northwest of Heathcote. It occurred near the boundary between Cambrian volcanic rocks (greenstones) and Silurian shales and sandstones and is thought to have been a lens of altered aluminium-rich rock within the greenstones. Extensive collecting has removed the exposure, which is now marked only by a pile of small greenstone pieces in a paddock on the NW corner of Coppermine Road and the Heathcote–Elmore Road. Similar rocks occur near Tatong and at Waratah Bay, also associated with Cambrian greenstones, but these occurrences have not been studied.

TURQUOISE

This attractive blue gem mineral has a history of use in jewellery and ornaments stretching back over 5000 years, with important sources in ancient Persia, Egypt and southwestern USA. In Victoria, veins of turquoise were discovered in the parish of Edi, about 45 km northeast of Mansfield, in the late 1880s. By 1893, when Edward Dunn visited the region, turquoise was being mined at eight sites spread over a distance of about 5 km (Dunn, 1908). The host rocks for the turquoise veins are black slates of Ordovician age which are exposed in the Black Range, east of the settlement of Cheshunt (Howitt, 1906). The veins are up to 2 cm thick but usually less than 1 mm, with the colour varying from pale or greenish-blue to intense sky-blue. According to a report by Hyman Herman in 1912, the best turquoise was an exquisite blue equal to the best Persian stone. He reported there was strong demand for the mineral in jewellery and that considerable quantities had been sold in England and Germany for ornamental inlays and cameos. However, little is known about how much turquoise was mined from the area and to whom it was sold. Nor are there any known surviving jewellery or ornamental items featuring Victorian turquoise. The main leaseholder, P. C. Gascoigne, endeavoured to establish a market

◊

9.5 Rough turquoise vein in slate with cabochons to 9 mm from near Cheshunt.

for the turquoise amongst local jewellers but it appears he was unsuccessful and mining ceased in about 1921. It is still possible to collect turquoise pieces on dumps at the main workings.

While the Edi–Cheshunt turquoise field is the most important occurrence in Victoria, turquoise has been recorded at other localities. These include Greta South (Howitt, 1906); near Licola and the Tara Range in east Gippsland (Teale, 1920), and on the Mornington Peninsula. More detailed descriptions of the history, geology and minerals of all these localities, including the Edi–Cheshunt field, appear in the book *Phosphate Minerals of Victoria* (Birch *et al.*, 1993).

Gem materials in other Australian states

Australian localities provide a wide range of 'miscellaneous' ornamental rocks and minerals from diverse geological environments. Some well-known examples are the bright pink rhodonite deposits in the Tamworth region of New South Wales; the prehnite occurrences at Prospect, New South Wales and on Wave Hill Station in the Northern Territory; the nephrite deposits at Cowell in South Australia; turquoise from Amaroo Station, Northern Territory; and stichtite from Dundas, Tasmania.

COWELL

Pods of dark green to black 'jade' are enclosed in dolomitic marble and calc-silicate rocks of Proterozoic age near Cowell, on the Eyre Peninsula of South Australia. Discovered and identified in the 1960s, the so-called jade is the variety known as nephrite, which at Cowell is the amphibole mineral ferroactinolite. Most of the jade is dark green, with a small proportion being black, which is the most valuable variety. The material is very fine-grained, so it takes a high polish. The deposit was mined on demand and a wide variety of decorative objects were carved and polished for export, but the current status of the operations is uncertain.

9.6 Polished and shaped nephrite ('jade') from Cowell, South Australia (pieces are 9 cm across).

WAVE HILL STATION (KALKARINGI)

Semi-precious prehnite occurs in basalt (the Early Cambrian Antrim Plateau Volcanics) on Wave Hill station near Kalkaringi, some 370 km SSW of Katherine in the Northern Territory. Cavities or geodes in the basalt are filled with prehnite, forming compact, yellowish-green crystalline aggregates up to 10 cm thick. These can be collected as nodules on the surface after they have weathered out of the enclosing rock. The translucent varieties of prehnite are the most sought-after by collectors for producing carvings and cabochons.

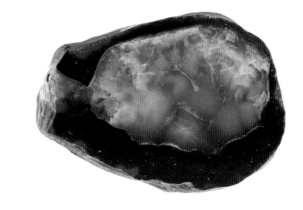

9.7 Polished prehnite nodule 12 cm across from Wave Hill station, Northern Territory.

9.8 Polished prehnite pieces up to 4 cm across from Wave Hill station.

9.9 *(below)* Faceted prehnite (10 x 15 mm) from Wave Hill Station.

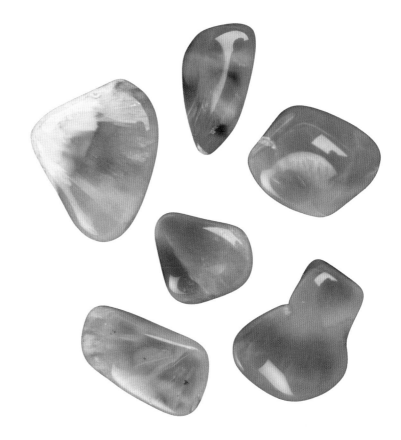

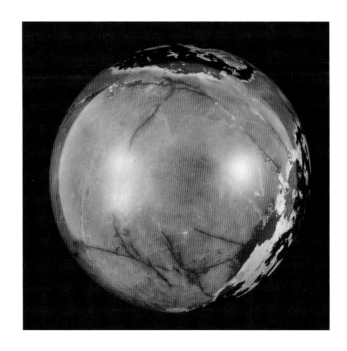

◇

(clockwise from above)

9.10 Deep pink rhodonite
(10 cm across) from the
Tamworth district,
New South Wales.

9.11 Polished sphere
(6 cm across) of rhodonite
from Tamworth district,
New South Wales.

9.12 Polished fragments of
rhodonite (up to 4 cm long)
from the Tamworth district,
New South Wales.

TAMWORTH DISTRICT

The finest deep-pink ornamental rhodonite, manganese silicate, comes from numerous small deposits north and east of Tamworth, in the southern New England region of New South Wales. Originally opened up for manganese oxides, several deposits are now mined sporadically on a small scale for rhodonite, which forms lenses enclosed in metamorphosed and silicified Silurian to Carboniferous sedimentary rocks. The rhodonite is massive and fine-grained, usually opaque, and shows colours ranging from cream to pale and bright pink; deep reddish-pink rhodonite is the most prized variety but is quite rare.

DUNDAS STICHTITE

Stichtite is a rare magnesium chromium carbonate with an intense purple colour. It occurs as patches and seams in green serpentine at Stichtite Hill, in the old Dundas mining field, western Tasmania. The attractive rock can be shaped and polished to make a variety of ornaments.

HARTS RANGE CORDIERITE

Cordierite, magnesium aluminium silicate, is widespread in high-grade metamorphic rocks throughout Australia. However, gem-quality violet-blue crystals, referred to as iolite, are found in only a few remote localities. Probably the best material comes from several sites in the Entire Valley in the Harts Ranges of Northern Territory. Crystalline masses up to 10 cm or more across occur in quartz-feldspar-mica schists, and have been recovered from shallow workings. While highly fractured, the crystals provide attractive gemstones up to several carats in weight.

◊

9.13 Polished sphere (6 cm across) of serpentine with purple stichtite from Dundas, Tasmania.

9.14 Faceted cordierite from the Harts Range, Northern Territory (7 mm).

References

Adams, H. and Krantz, D. (1983): Amethyst crystals from Springhurst. *Australian Gem and Treasure Hunter*, 85, 13.

Amess, A. (1973): Cutting the Crystal King. *Australian Lapidary Magazine*, June, 17–20.

Anderson, C. A. (1908): Mineralogical Notes: No. VI. Topaz, beryl, anglesite, rutile, atacamite, pyrite. *Records of the Australian Museum*, 7, 60–68.

Anon. (1854): Official Catalogue of the Melbourne Exhibition, 1854, in connection with the Paris Exhibition, 1855. F. Sinnett and Company, Melbourne.

Anon. (1869): A Supplementary Catalogue of Minerals, Rocks and Fossils which have been collected by the Mining Department, Melbourne, Victoria. Government Printer, Melbourne.

Anon. (1872): Official Catalogue of Exhibits, London Exhibition 1873 - Melbourne Exhibition 1872-3. Published by the Exhibition Commissioners. Masson Firth and McCutcheon, Melbourne.

Anon. (1873): The Vienna Universal Exhibition of 1873, Victoria, Australia. Official Catalogue of Exhibits.

Anon. (1886): Victorian Court Catalogue of Exhibits, Colonial and Indian Exhibition, London 1886, 185–201.

Anon. (1966a): Victorian gemstone locations: Moonlight Head. *The Australian Gemhunter*, October, 4–6.

Anon. (1966b): Victorian gemstone locations: Crystal King mine, Tallangalook. *The Australian Gemhunter*, November, 3–5.

Anon. (1975): Giant agate find. *The Lapidarian*, February, 8.

Atkinson, J. A. (1897): A locality list of all the minerals hitherto recorded from Victoria. *Proceedings of the Royal Society of Victoria*, 9, 68–119.

Baker, G. (1955): Australites from Harrow, Victoria. *Mineralogical Magazine*, 30 (228), 596–603.

Baker, G. (1962): Detrital heavy minerals in natural accumulates, with special references to Australian occurrences. Monograph Series No. 1, Australasian Institute of Mining and Metallurgy, Melbourne.

Baragwanath, W. (1925a): Cassiterite at Beenak. *Records of the Geological Survey of Victoria*, 4(4), 415.

Baragwanath, W. (1925b): The Aberfeldy District, Gippsland. *Memoirs of the Geological Survey of Victoria*, No. 15.

Baragwanath, W. (1948): Diamonds in Victoria. *Mining and Geological Journal*, 3(3), 12–16. Department of Mines, Victoria.

Barron, L. M., Lishmund, S. R., Oakes, G. M., Barron, B. J. and Sutherland, F. L. (1996): Subduction model for the origin of some diamonds in the Phanerozoic of eastern New South Wales. *Australian Journal of Earth Sciences*, 43, 257–267.

Beasley, A.W. (1966): Olivine in Victorian volcanic bombs. *The Australian Gemmologist*, 64, 97-99.

Beasley, A. W. (1970): Unusually large specimens of topaz and augite. *Victorian Naturalist*, February, 1–3.

Beasley, A. W. (1978): Minerals and rocks on Phillip Island. *Australian Gems and Crafts Magazine*, 29, 11–12.

Beasley, A. W. (1979): Gemstones on Australian beaches. *Australian Gems and Crafts Magazine*, 40, 9–11.

Becker, L. (1857): Victorian Minerals. *Neues Jahrbuch fur Mineralogie*, 312–315.

Bell, G. (1953): Beach sand from Betka River near Mallacoota. *Unpublished Reports of the Geological Survey of Victoria*, 1953/24.

Birch, W. D. (1984): Quartz-topaz-loellingite rocks near Eldorado, Victoria. *Australian Journal of Earth Sciences*, 31, 269–278.

Birch, W. D. (1986): Gemstones of the Beechworth area. *Australian Gemmologist*, 16(3), 101–106.

Birch, W. D. (1993): Phosphate minerals in granitic rocks. In Birch, W. D. and Henry, D. A. (Editors), *Phosphate Minerals of Victoria*, 3–39. Special Publication No. 3, Mineralogical Society of Victoria, Melbourne.

Birch, W. D., Henry, D. A., Leach, J., Robinson, F., Haupt, J., Day, B. and Bosworth, J. (1993): Sedimentary phosphate deposits. In Birch, W. D. and Henry, D. A. (Editors), *Phosphate Minerals of Victoria*, 65–110. Special Publication No. 3, Mineralogical Society of Victoria, Melbourne.

Birch, W. D. (1996): Danburite-bearing calc-silicate rocks from the Ascot Hills, Dookie, Victoria. *Australian Journal of Earth Sciences*, 43, 387–393.

Birch, W. D. (2007): Gemstones in the forest — Rev. John Bleasdale's journey to William Wallace Creek, Victoria. In: Pierson, R. R. (ed.) *The history of geology in the second half of the nineteenth century; the story in Australia, and in Victoria, from Selwyn and McCoy to Gregory – 1853 to 1903.* Earth Science History Group Conference, December 2007. Volume of Short Papers, Special Publication No. 1, Earth Sciences History Group,

GSA Inc., Melbourne, Victoria. Pp 8–10.

Birch, W. D. (2008): Gem corundum from the St Arnaud district, western Victoria, Australia. *Australian Journal of Mineralogy*, 14(2), 25–30.

Birch, W. D., Magee, C., Barron, L. M. and Sutherland, F. L. (2007): Gold- and diamond-bearing White Hills Gravel, St Arnaud district, Victoria: age and provenance based on U-Pb dating of zircon and rutile. *Australian Journal of Earth Sciences*, 54(4), 609–628.

Blake, L. J. (1977): *Place Names of Victoria*. Rigby Limited.

Bleasdale, J. I. (1865): On Precious Stones. *Transactions of the Royal Society of Victoria*, vi, 98–104.

Bleasdale, J. I (1866a): A Report on the Results of an Exhibition of Gems, etc. *Transactions of the Royal Society of Victoria*, vii, 64–92.

Bleasdale, J. I. (1866b): Gems and Sapphires. *Transactions of the Royal Society of Victoria*, vii, 147–149.

Bleasdale, J. I. (1866c): On Gems and Precious Stones found in Victoria. Intercolonial Exhibition Essays, 1866 (1867), No. 4, (pp. 237–248 of Official Record).

Bleasdale, J. I. (1868a): Colonial Gem Stones. The Corundum Series. *Colonial Monthly*, vol. 3, No. 14 (October 1868), 137–143.

Bleasdale, J. I. (1868b): On Rubellite — Red Tourmaline — found at Tarrengower, Victoria. *Transactions of the Royal Society of Victoria*, viii, 225–227.

Bleasdale, J. I. (1869): On Colonial Gems. *Transactions of the Royal Society of Victoria*, ix, 72–75.

Bleasdale, J. I. (1876): Victorian Gems and Precious Stones. Philadelphia Centennial Exhibition, 1876 (Melbourne, 1875), Catalogue of Exhibits, Victorian Department, 71–75.

Bolger, P. (1977): Explanatory notes on the Meredith and You Yangs 1:50 000 geological maps. *Reports of the Geological Survey of Victoria*, 1977/14.

Boutakoff, N. and Whitehead, S. (1952): Enhydros or waterstones. *Mining and Geological Journal*, 4(5), 14–19. Department of Mines, Victoria.

Butterworth, M. (1977): Thundereggs at the end of the rainbow. *Australian Gems and Crafts Magazine*, 23, 109–111.

Canavan, F. (1988): Deep Lead Gold Deposits of Victoria. *Bulletins of the Geological Survey of Victoria*, No. 62.

Chalmers, R. O. (1967): *Australian Rocks, Minerals and Gemstones*. Angus & Robertson Ltd., Sydney.

Clark, D. (1892): Minerals of East Gippsland. Australasian Association for the Advancement of Science, 4, 285–289.

Cochrane, G. W. (1971): Tin Deposits of Victoria. *Bulletins of the Geological Survey of Victoria*, No. 60.

Coldham, T. (1985): Sapphires from Australia. *Gems and Gemology*, 21, 130–146.

Cooksey, T. (1895): Some suggestions regarding the formation of enhydros or waterstones. *Records of the Australian Museum*, 2(6), 92–94.

Couchman, T. (1876): Mineral Statistics for 1875, Victorian Department of Mines. Parliamentary Paper, Victoria, No. 12.

Couchman, T. (1877): Mineral Statistics for 1876, Victorian Department of Mines. Parliamentary Papers, Victoria, No. 6.

Couchman, T. (1878): Mineral Statistics for 1877, Victorian Department of Mines. Parliamentary Papers, Victoria, No. 8.

Couchman, T. (1879): Mineral Statitstics for 1878, Victorian Department of Mines. Parliamentary Papers, Victoria, No. 5.

Couchman, T. (1880): Mineral Statistics for 1879, Victorian Department of Mines. Parliamentary Papers, Victoria, No. 10.

Couchman, T. (1881): Mineral Statistics for 1880, Victorian Department of Mines. Parliamentary Papers, Victoria, No. 80.

Couchman, T. (1882): Mineral Statistics for 1881, Victorian Department of Mines. Parliamentary Papers, Victoria, No. 29.

Couchman, T. (1883): Mineral Statistics for 1882, Victorian Department of Mines. Parliamentary Papers, Victoria, No. 3.

Coulson, A. (1954): The volcanic rocks of the Daylesford district. *Proceedings of the Royal Society of Victoria*, 65, 113–124.

Crohn, P. W. (1952): Piezo-electric quartz at the Crystal King mine, Tallangallook. *Mining and Geological Journal*, 4(5), 29–33. Department of Mines, Victoria.

Cutter, I. A. (1967): Crystal King mine; how to get there. *The Australian Gemhunter*, February, 12-14.

Day, R. A. (1978): A description of the Meredith ultramafic breccia. Appendix 1. In Hunt, F. L., Exploration possibilities for diamonds in Victoria. Unpublished report 1978/91, Department of Minerals and Energy, Victoria.

Day, R. A., Nicholls, I. A. and Hunt, F. L. 1979: The Meredith ultramafic breccia pipe — Victoria's first kimberlite? In Denham, D (Editor), Crust and Upper Mantle of Southeast Australia. *Records of the Bureau of Mineral Resources, Australia*, 1979(2), 22–24.

Dineen, H. (1969): Snowy River Rocks. *Australian Lapidary Magazine*, June, 17–20.

Dineen, H. (1981): Moonlight Head. *The Lapidarian*, June, 7–11.

Dunn, E. J. (1870): On the occurrence of enhydros or waterstones at Beechworth. *Proceedings of the Royal Society of Victoria*, 10(1), 32–35.

Dunn, E. J. (1871): Notes on the Rocks and Minerals of the Ovens District. *Reports of the Mining Surveyors and Registrars, Victoria*, No. 26, March 31, 41–47.

Dunn, E. J. (1887): Gold Mining in the Chiltern District. *Quarterly Reports of the Mining Surveyors and Registrars, Victoria*. December, Appendix F, 84–85.

Dunn, E. J. (1890): Notes on the Rheola Gold-field. *Quarterly Reports of the Mining Surveyors and Registrars, Victoria*. June, 17–23.

Dunn, E. J. (1907): The Morrison Gold-Field. *Records of the Geological Survey of Victoria*, 2(1), 24–25.

Dunn, E. J. (1907): Red jasper in the Heathcotian rocks, near Tooleen. *Records of the Geological Survey of Victoria*, 2(2), 81.

Dunn, E. J. (1908a): The Edi turquoise field, King River. *Records of the Geological Survey of Victoria*, 2(4), 170–175.

Dunn, E. J. (1908b): The general features of the Jamieson Goldfield and some mines between Jamieson and Wood's Point. *Records of the Geological Survey of Victoria*, 3(1), 41–54.

Dunn, E. J. (1908c): Wolframite deposits near Marysville. *Records of the Geological Survey of Victoria*, 3(2), 259–260.

Dunn, E. J. (1909a): The Chiltern Valley Gold Mine, No. 2 Shaft, and the Golden Bar Mine, Chiltern. *Records of the Geological Survey of Victoria*, 3(1), 56–59.

Dunn, E. J. (1909b): The serpentine area, Wellington River, Gippsland. *Records of the Geological Survey of Victoria*, 3(1), 65–68.

Dunn, E. J. (1913): The Woolshed Valley, Beechworth. *Bulletins of the Geological Survey of Victoria*, No. 25.

Dunn, E. J. (1917): Mines at Tallangalook and Bonnie Doon. *Records of the Geological Survey of Victoria*, 4(1), 66–69.

Edwards, Dianne H. (1979): The Diamond Valley Story. Shire of Diamond Valley, Greensborough.

Ferguson, W. H. (1897): Report on the Middle Creek and Hoorigans Creek Goldfields. *Unpublished Reports of the Geological Survey of Victoria*, 1897/7118.

Ferguson, W. H. (1902): Notes on common opal deposits in East Gippsland. *Records of the Geological Survey of Victoria*, 1(1), 85.

Ferguson, W. H. (1906a): The Blackwood–Trentham Gold-field. *Bulletins of the Geological Survey of Victoria*, No. 18.

Ferguson, W. H. (1906b): Report on the glacial conglomerate of supposed Jurassic age in Parish of Wonga Wonga, near Foster, southern Gippsland. *Records of the Geological Survey of Victoria*, 1(4), 249–251.

Ferguson, W. H. (1917): The discovery of fossils in the Grampians Sandstones. *Records of the Geological Survey of Victoria*, 4, 5–9.

Ferguson, W. H. (1936): Turtons Creek Goldfield. *Records of the Geological Survey of Victoria*, 5(2), 249–252.

Flett, J. (1970): *The history of gold discovery in Victoria*. The Hawthorn Press, Melbourne.

Foord, G. (1862): On the Occurrence of the Diamond and Chloro-bromide of Silver amongst the Gold Ores of Australia. *Chemical News*, vi, 14.

Foord, G. (1871): Notes on enhydros found at Beechworth. *Proceedings of the Royal Society of Victoria*, x, 71–74.

Grayson, A. J. and Mahony, D. J. (1910): The geology of the Camperdown and Mount Elephant districts. Memoirs of the Geological Survey of Victoria, 9.

Hall, T. S. (1895): The geology of Castlemaine, with a subdivision of part of the Lower Silurian rocks of Victoria, and a list of Minerals. *Proceedings of the Royal Society of Victoria*, 7, 55–88.

Hall, A. E. and Smith, C. B. (1982): Morphological study of diamonds held by the Victorian Museum. C.R.A. Exploration Report 130100 (unpublished).

Harris, W. J. and Thomas, D. E. (1948): The geology of Campbelltown. Mining and Geological Journal, 3(3), 46–54. Department of Mines, Victoria.

Hart, T. S. (1905): The mineralogical character of Victorian auriferous occurrences. *Proceedings of the Royal Society of Victoria*, 18, 25–37.

Herman, H. (1912): Victorian Minerals. Annual Report to the Secretary of Mines, Victoria, 117.

Herman, H. (1914): Economic geology and mineral resources of Victoria. *Bulletins of the Geological Survey of Victoria*, 34.

Hocking, J. B., Gloe, C. S., Threlfall, W. F., Holdgate, G. R. and Bolger, P. F. (1988): Gippsland Basin, in Douglas, J. G. and Ferguson, J. A. (Editors), *Geology of Victoria*. Geological Society of Australia, Victorian Division.

Hollis, J. D., Sutherland, F. L. and Gleadow, A. J. W. (1986): The occurrence and possible origins of large zircons in alkali volcanics of eastern Australia. In *Crystal Chemistry of Minerals*, 565–578. Publishing House of the Bulgarian Academy of Science, Sophia.

Hollis, J. D. and Sutherland, F. L. (1989): Gem zircon, corundum and other heavy minerals from north central Victoria: Ages of volcanism, uplift and drainage systems. Collected Abstracts, R. O Chalmers Symposium, Australian Museum, Sydney.

Howard, R T. (1972): Acheron Cauldron and Black Range. BSc.Hons Thesis (unpublished), University of Melbourne.

Howitt, A. M. (1906a). Report on the Edi-Myrrhee turquoise belt and the chert and jasper belts near Tatong. *Records of the Geological Survey of Victoria*, 1(4), 239–241.

Howitt, A. M. (1906b): Report on lignite deposits near Mahaikah, County of Delatite. *Records of the Geological Survey of Victoria*, 1(4), 241–243.

Howitt, A. M. (1921): Wolfram at Britannia Creek. *Records of the Geological Survey of Victoria*, 4(3), 265–268.

Howitt, A. W. (1898): On oligoclase feldspar from Mount Anakies in Victoria. *Reports of the Australasian Association for the Advancement of Science*, vii, 375–377.

Hunt, F. L. (1978): Exploration possibilities for diamonds in Victoria. *Unpublished Report of the Department of Minerals and Energy*, Victoria, 1978/91.

Hunter, S. B. (1909): The Deep Leads of Victoria. *Memoirs of the Geological Survey of Victoria*, No. 7.

Irving, A. J. (1974): Megacrysts from the Newer Basalts and other basaltic rocks of southeastern Australia. *Bulletin of the Geological Society of America*, 85, 1503–1514. Jacobson, R. and Scott, T.R. (1937): The geology of the Korkuperrimul Creek area, Bacchus Marsh. *Proceedings of the Royal Society of Victoria*, 50, 110–155.

Junner, N. R. (1914): The petrology of the igneous rocks near Healesville and Narbethong. *Proceedings of the Royal Society of Victoria*, 27(1), 261–185.

Jutson, J. T. (1905): Notes on the volcanic history of Mt Shadwell. *Victorian Naturalist*, 22, 8–12 (with Appendix by Chapman, F. on some rocks and minerals from Mount Shadwell).

Keble, R. A. (1950): The Mornington Peninsula. *Memoirs of the Geological Survey of Victoria*, No. 17.

King, R. (1985): Explanatory notes on the Ballarat 1:250 000 geological map. *Reports of the Geological Survey of Victoria*, No. 75.

Kitson, A. E. (1896): Geological notes on the Toombullup Goldfield and adjacent country. *The Victorian Naturalist*, 107–111.

Kitson, A. E. (1901): Toora district. *Unpublished Reports of the Geological Survey of Victoria*, 1901/-.

Kitson, A. E. (1902a): Report on the probable occurrence of coal in the Welshpool district. *Records of the Geological Survey of Victoria*, 1(1), 61–66.

Kitson, A. E. (1902b): Report on an alleged coal seam near Ruby, and on the geology of that locality. *Unpublished Reports of the Geological Survey of Victoria*, 1902/4888.

Kitson, A. E. (1903a): Glacial deposits of Taminick, Glenrowan and Greta, north-eastern Victoria. *Proceedings of the Royal Society of Victoria*, 16, 148–153.

Kitson, A. E. (1903b): Volcanic necks at Andersons Inlet, South Gippsland, Victoria. *Proceedings of the Royal Society of Victoria*, 16, 154–177.

Kitson, A. E. (1905): The economic rocks and minerals of Victoria. Victorian Year-book for 1905, Part IX, 517–536.

Kitson, A. E. (1917): The Jumbunna and Powlett Plains district, South Gippsland. *Bulletins of the Geological Survey of Victoria*, 40, 1–26.

Krause, F. M. (1896): An Introduction to the Study of minerals for Australian Readers. George Robertson and Company, Melbourne.

Langtree, C. W. (1883): Mineral Statistics for 1882, Victorian Department of Mines. Parliamentary Papers, Victoria, No. 37.

Lavell, L., Butterworth, M. and Butterworth, G. (1969): Rockhunting in the cave country of Buchan. James Yeates and Sons, Bairnsdale.

Louthean, R. (1991): Register of Australian Mining 1991/92. Vanguard Press, Perth.

McKenzie, D. A., Nott, R. J. and Bolger, P. F. (1984): Radiometric age determinations. *Reports of the Geological Survey of Victoria*, No. 74.

McMahon, B. and Hollis, J. D. (1983): Zircons from Lake Bullenmerri. Victoria. Australian Mineralogist, 42, 235–236.

McMillan, A. (1864): Journal of the Alpine Expedition (Gippsland). In State Library of Victoria (copy). MS 9776 (on microfilm).

Macumber, P. (1978): Permian glacial deposits, tectonism, and the evolution of the Loddon Valley. *Mining, Geology and Energy Journal of Victoria*, 7(3), 34–36. Department of Minerals and Energy, Victoria.

Mahony, D. J. (1922): The mineral composition of selwynite. *Records of the Geological Survey of Victoria*, 4(4), 474–475.

Mahony, D. J. (1928): Some Tertiary volcanic minerals and their parent magma. *Proceedings of the Royal Society of Victoria*, 40, 62–69.

Marsden, M. A. H. (1988): Upper Devonian – Carboniferous. In Douglas, J. G. and Ferguson, J. A. (Editors). *Geology of Victoria*. Geological Society of Australia, Victorian Division.

Menzies, M. A. (1995): Topaz. *The Mineralogical Record*, 26(1), 5–53.

Moon, R. A. (1897): Report on the Maldon Goldfield. Special Report of the Mines Department, Victoria.

Morand, V. J., Wohlt, K. E., Cayley, R. A., Taylor, D. H., Kemp, A. I. S., Simons, B. A. and Magart, A. P. M. (2003): Glenelg special map area. *Geological Survey of Victoria Report 123*. Department of Primary Industries.

Morvell, G. L. (1976): Mineral occurrences at Dookie, Victoria. *Australian Mineralogist*, 6, 21–23.

Mulvany, J. B. (1939): The Guildford Plateau Deep Lead Mine. *Mining and Geological Journal*, 2(1), 7–10.

Murphy, J. (1988): *No Parallel, the Woorayl Shire, 1888–1988*. Shire of Woorayl and Hargreen Publishing, North Melbourne.

Murray, R. A. F. (1876): Report on the Geology and Mineral Resources of South Western Gippsland. *Progress Reports of the Geological Survey of Victoria*, 3, 134–173.

Murray, R. A. F. (1884). Report of the examination of the country around Gembrook. *Progress Reports of the Geological Survey of Victoria*, 7, 18–19.

Murray, R. A. F. (1890): Report on the Franklin and Agnes River Tinfield. *Quarterly Reports of the Mines Department, Victoria*. September, 27–29.

Murray, R. A. F. (1891): Report on the Discovery of Stream Tin Ore near the Bunyip River. *Quarterly Reports of the Mines Department, Victoria*, September, 20.

Murray, R. A. F. (1914): The Mount William or Mafeking Goldfield. *Bulletins of the Geological Survey of Victoria*, 32.

Myatt, B. (1972): *Australian and New Zealand Gemstones*. Paul Hamlyn, Sydney, New South Wales.

Newbery, J. C. (1868): On the ornamental stones of the Colony. *Proceedings of the Royal Society of Victoria*, 9(2), 79–85.

Newbery, J. C. (1870): [Laboratory Report]. Mineral Statistics for 1869. Parliamentary Papers, Volume ii, Appendix D, 50–52.

Nicholas, W. (1876): Localities of Minerals which occur in Victoria. *Progress Reports of the Geological Survey of Victoria*, 3, 280–288.

Oakes, G. M., Barron, L. M. and Lishmund, S. R. (1996); Alkali basalts and associated volcaniclastic rocks as a source of sapphire in eastern Australia. *Australian Journal of Earth Sciences*, 43(3), 289–298.

Officer, G. and Hogg, Evelyn G. (1897): The Geology of Coimadai. *Proceedings of the Royal Society of Victoria*, 10(1), 60–74.

O'Reilly, S. Y., Nicholls, I. A. and Griffin, W. L. (1989): Xenoliths and megacrysts of Eastern Australia. In Johnston, R. W. (Ed.), *Intraplate volcanism in Eastern Australia and New Zealand*. Cambridge University Press.

Owen, H. B. (1945): Report on the Crystal King mine, Tallangalook, near Mansfield, Victoria. *Unpublished Report of the Geological Survey of Victoria*, 1945/13, 1–12.

Pauley, G. (1966): Victorian Gemstone Locations: Rysons Creek – The Bunyip River District. *The Australian Gemhunter*, December 1966, 3–5.

Pauley, G. R. and Smith, J. W. (1967): Victorian Gemstone Localities: Point Leo. *The Australian Gemhunter*, March 1967, 16.

Poth, R. (1984a): Diamonds in Victoria. *Australian Gem and Treasure Hunter*, 96, 24.

Poth, R. (1984b): Beechworth citrine. *Australian Gem and Treasure Hunter*, 92, 41.

Price, C. (1979): The Andersons Gully quartz. *Australian Gems and Crafts Magazine*, 39, 35.

Rickards, R. D. (1991): CRA Exploration Pty Ltd. EL 2558 Beatons, Victoria. Final and six monthly report for the period ending 4 July, 1991. Department of Energy and Minerals, Victoria, Expired Mineral Exploration Reports File (unpublished).

Rossiter, A. G. (2003). Granitic rocks of the Lachlan Fold Belt in Victoria. In: Birch, W. D. ed. *Geology of Victoria*, pp. 217–237. Geological Society of Australia Special Publication 23. Geological Society of Australia (Victoria Division).

Saxton, J. G. (1907): *Victoria place-names and their origin*. Saxton and Buckle, Printers and Publishers, Clifton Hill (Melbourne).

Scurfield, G. and Segnit, E. R. (1984): Petrifaction of wood by silica minerals. *Sedimentary Geology*, 39, 149–167.

Selwyn, A. R. C. (1861a): Catalogue of the Victorian Exhibition. Government Printer, Melbourne.

Selwyn, A. R. C. (1861b): Reports relating to the Geological Survey of Victoria. Votes and Proceedings of the Legislative Assembly, 1861-2, Volume 1.

Selwyn, A. R. C., Ulrich, G. H. F., Aplin, C. D. H., Etheridge, R. and Taylor, N. (1868): *A Descriptive Catalogue of the Rock Specimens and Minerals in the National Museum, collected by the Geological Survey of Victoria.* Government Printer, Melbourne.

Skeats, E. W. (1908): On the evidence of the origin, age and alteration of the rocks near Heathcote. *Proceedings of the Royal Society of Victoria*, 21(1), 302–348.

Smith, J. W. (1967a): Victorian Gemstone Locations; Cardinia Creek. *The Australian Gemhunter*, February, 6–7.

Smith, J. W. (1967b): Victorian Gemstone Locations; Kitty Miller Bay. *The Australian Gemhunter*, April, 6–8.

Smyth, R. B. (1865): Mineral Statistics for 1864, Victorian Department of Mines. Parliamentary Papers, Victoria, No. 41.

Smyth, R. B. (1866a): Mineral Statistics for 1865, Victorian Department of Mines. Parliamentary Papers, Victoria, No. 4.

Smyth, R. B. (1866b): A Catalogue of Minerals, Rocks and Fossils which have been collected in the Colony by the Mining Department, Melbourne, Victoria. Government Printer, Melbourne.

Smyth, R. B. (1867): Mineral Statistics for 1866, Victorian Department of Mines. Parliamentary Papers, Victoria, No. 43.

Smyth, R. B. (1868): Mineral Statistics for 1867, Victorian Department of Mines. Parliamentary Papers, Victoria, No. 11.

Smyth, R. B. (1869a): *The Goldfields and Mineral Districts of Victoria.* Government Printer, Melbourne.

Smyth, R. B. (1869b): *A supplementary catalogue of minerals, rocks and fossils which have been collected by the Mining Department, Melbourne, Victoria.* Government Printer, Melbourne.

Smyth, R. B. (1869c): Mineral Statistics for 1868, Victorian Department of Mines. Parliamentary Papers, Victoria, No. 39.

Smyth, R. B. (1870): Mineral Statistics for 1869, Victorian Department of Mines. Parliamentary Papers, Victoria, No. 4.

Smyth, R. B. (1871): Mineral Statistics for 1870, Victorian Department of Mines. Parliamentary Papers, Victoria, No. 8.

Smyth, R. B. (1872a): Mineral Statistics for 1871, Victorian Department of Mines. Parliamentary Papers, Victoria, No. 8.

Smyth, R. B. (1872b): Victorian Exhibition, 1872. Mining & mineral statistics; with notes on the rock formations of Victoria. Mason, Firth and McCutcheon, Melbourne.

Smyth, R. B. (1873): Mineral Statistics for 1872, Victorian Department of Mines. Parliamentary Papers, Victoria, No. 7.

Smyth, R. B. (1874): Mineral Statistics for 1873, Victorian Department of Mines. Parliamentary Papers, Victoria, No. 8.

Smyth, R. B. (1875): Mineral Statistics for 1874, Victorian Department of Mines. Parliamentary Papers, Victoria, No. 2.

Spencer, L. (1970): Maldon and minerals. *Australian Lapidary Magazine*, 7(2), 3–9.

Spencer-Jones, D. (1955): Geology of the Toora Tin-field. *Bulletins of the Geological Survey of Victoria*, No. 54.

Stephen, G. M. (1854): On the gems and gold crystals of the Australian Colonies. *Quarterly Journal of the Geological Society of London*, 303–308.

Stirling, J. 1891: Report on gold discoveries, Crichton's Creek, Gembrook. Reports and Statistics of the Mining Department, June 30.

Stirling, J. (1894): Notes on the Foster Gold-field and District. *Progress Reports of the Geological Survey of Victoria*, 8, 66.

Stirling, J. (1898a): Report on Toombullup Gold-field. *Progress Reports of the Geological Survey of Victoria*, 9, 45–46.

Stirling, J. (1898b): Report on the geological and mining features of portion of the Western District. *Progress Reports of the Geological Survey of Victoria*, 9, 86–91.

Stirling, J. (1899a): Geological sketch plan and section of the Werribee Gorge, near Bacchus Marsh. *Monthly Progress Reports of the Geological Survey of Victoria*, No. 2, 25.

Stirling, J. (1899b): Further Report on Gembrook District (Cockatoo Creek). Notes on the Rocks of the Gembrook District. *Monthly Progress Reports of the Geological Survey of Victoria*, 8 & 9, 6–8.

Stirling, J. (1899c): Report on gold workings, Stony Creek, Gippsland. *Monthly Progress Reports of the Geological Survey of Victoria*, 8 & 9, 21–22.

Stirling, J. (1899d): Notes on the Bullengarook Plateau, 9 miles north of Bacchus Marsh. *Monthly Progress Reports of the Geological Survey of Victoria*, 8 & 9, 49.

Stone, D. (1967): *Gemstones of Victoria.* Jacaranda Press Pty Ltd.

Sutherland, F. L. (1996): Alkaline rocks and gemstones, Australia: a review and synthesis. *Australian Journal of Earth Sciences*, 43(3), 323–343.

Sutherland, F. L. and Fanning, C. M. (2001): Gem-bearing basaltic volcanism, Barrington, New South Wales: Cenozoic evolution, based on basalt K–Ar ages and zircon fission track and U–Pb isotope dating. *Australian Journal of Earth Sciences*, 48, 221–337.

Sutherland, F. L. and Schwarz, D. (2001): Origin of gem corundums from basaltic fields. *Australian Gemmologist*, 21, 30–33.

Sutherland, F. L., Schwarz, D., Jobbins, E. A., Coenraads, R. R. and Webb, G. (1998): Distinctive gem corundum suites from basaltic fields: a comparative study of Barrington, Australia, and West Pailin, Cambodia, gemfields. *Journal of Gemmology*, 26, 65–85.

Sutherland, F. L., Zaw, K., Meffre, S., Giuliani, G., Fallick, A. E., Graham, I. T. and Webb, G. B. (2009): Gem-corundum megacrysts from east Australian basalt fields: trace elements, oxygen isotopes and origins. *Australian Journal of Earth Sciences*, 56, 1003–1022.

Sutton, P. and Roberts, D. (2010): Quartz and associated minerals from the Pilot Range, Beechworth, Victoria. *Australian Journal of Mineralogy*, 16(1), 25–33.

Tan, S. H. (1982): Diamond potential of Victoria — a preliminary report. *Unpublished Reports of the Geological Survey of Victoria*, 1979/112.

Tangney, D. (1991): The source of diamonds at Beechworth, Victoria. Final Report, Geological Mapping Project (unpublished). Royal Melbourne Institute of Technology.

Taylor, W. R., Jaques, A. L. and Ridd, M. (1990): Nitrogen-defect aggregation characteristics of some Australian diamonds: Time-temperature constraints on the source regions of pipe and alluvial diamonds. *American Mineralogist*, 75, 1290–1310.

Teale, E. O. (1920): Palaeozoic Geology of Victoria. *Proceedings of the Royal Society of Victoria*, 32, 85–146.

Teichert, C. and Talent, J. A. (1958): Geology of the Buchan area, East Gippsland. *Memoirs of the Geological Survey of Victoria*, 21.

Thiele, E. O. (1908): Notes on the Dolodrook serpentine area and the Mt. Wellington Rhyolites, North Gippsland. *Proceedings of the Royal Society of Victoria*, 21(1), 249–269.

Tickell, S. J., Edwards, J. and Abele, C. (1992): Port Campbell embayment. *Reports of the Geological Survey of Victoria*, 95.

Towsey, C. A. J. (1979): Pyrope garnets from the Koetong district, north-eastern Victoria. *Australian Mineralogist*, 22, 105–106.

Ulrich, G. H. F. (1866): Mineral Species of Victoria. Intercolonial Exhibition Essays, No. 3, 1866-7 (pp.184–235 of Official Catalogue).

Ulrich, G. H. F. (1868): Notes on Quarter Sheet 14 NW (Bradford, Loddon District). Geological Survey of Victoria.

Ulrich, G. H. F. (1870): *Contributions to the Mineralogy of Victoria.* Mineral Statistics for 1869, Parliamentary Papers, Victoria, Appendix E, pp.52–67; also separate (Melbourne, 1870).

VandenBerg, A. H. M. (1971): Explanatory notes on the Ringwood 1:63 360 geological map, *Reports of the Geological Survey of Victoria*, 1971/1.

Walcott, R. H. (1901): Additions and Corrections to the Census of Victorian Minerals. *Proceedings of the Royal Society of Victoria*, xxxi (ii), 253–272.

Watson, J. C. (1922): The chemical composition of selwynite. *Records of the Geological Survey of Victoria*, 4(4), 472–474.

Wellman, P. (1974): Potassium–argon ages on the Cainozoic volcanic rocks of eastern Victoria, Australia. Journal of the Geological Society of Australia, 21(4), 359–376.

White, J. (1974): One hundred years of history; the Shires of Bass and Phillip Island. South Gippsland Central Times Publishing Company Pty Ltd.

Young, D. N. (1983): Geology and geochemistry of the granites of the Pilot Range, NE Victoria. BSc.(Hons.) thesis (unpublished), La Trobe University, Melbourne.

References for other Australian gemstone localities

Anon, (1986): A guide to fossicking in the Northern Territory (Second edition: Ed. D. Thompson). Northern Territory Department of Mines and Energy, 73pp.

Bottrill, R. S. And Baker, W. E. (2008): A catalogue of the minerals of Tasmania. *Bulletin Geological Survey Tasmania*, 73.

Chalmers, R. O. (1967): Australian rocks, minerals and gemstones. Angus & Robinson Ltd, Sydney. 398 pp.

Featherston, J. M., Stocklmayer, S. M. and Stocklmayer, V.C. (2013): *Gemstones of Western Australia.* Geological Survey of Western Australia, *Mineral Resources Bulletin* 25, 306p.

Grundmann, G. (1998): Alexandrite, emerald, ruby, sapphire and topaz in a biotite-phlogopite fels from Poona, Cue district, Western Australia. *The Australian Gemmologist*, 20, 159–167.

Jacobson, M I, Calderwood, M A. and Grguric, B A. 2007. Guidebook to the pegmatites of Western Australia. Hesperian Press, Carlisle, WA. 356 pp.

Sutherland, L. (1991): *Gemstones of the Southern Continents*. Reed Books Pty Ltd, New South Wales.

Webb, G. and Sutherland, F. L. (1998): Gemstones of New England. *Australian Journal of Mineralogy*, 4(2), 115–121.

Various papers on Australian gemstone occurrences and properties. See website for *The Australian Gemmologist* (Journal of the Gemmological Association of Australia) http://www.australiangemmologist.com.au/indices.html

CHAPTER 2: DIAMONDS
Ahmat, A. L. (2012): The Ellendale Diamond Field: exploration history, discovery, geology and mining. *The Australian Gemmologist*, 24(12), 280–288.

Bevan, A. and Downes, P. (2004): In the pink: Argyle's diamond gift to Australia. *The Australian Gemmologist*, 22(4), 150–155.

Downs, P. J., Bevan, A. W. R. and Deacon, G. L. (2012): The Kimberley Diamond Company Ellendale diamond collection at the Western Australian Museum. *The Australian Gemmologist*, 24(12), 289–293.

CHAPTER 3: SAPPHIRES, RUBIES & ZIRCONS
Abduriyim, A., Sutherland, F. L. and Coldham, T. (2012): Past, present and future of Australian gem corundum. *The Australian Gemmologist*, 24(10), 234–242.
Brown, G., Bracewell, H. and Snow, J. (1989): Gems of the Mud Tank Carbonatites. *The Australian Gemmologist*, 17(2), 52–57.

McColl, D. H. and Warren, R. G. (1979): The first discovery of ruby in Australia. *Australian Mineralogist*, 26, 121–125.

Webb, G. (2007): Ruby suites from New South Wales. *The Australian Gemmologist*, 23(3), 99–116.

CHAPTER 4: OLIVINE & ANORTHOCLASE
Beasley, A. W. (1974): Peridot in Eastern Australia. *Australian Gems and Crafts Magazine*, 8, 15.

Bracewell, H. (2000): Rare Australian gemstones; moonstone, a rare Queensland gemstone. *The Australian Gemmologist*, 20(12), 523–528.

CHAPTER 5: TOPAZ, TOURMALINE & BERYL
Bottrill, R. S., Woolley, R. N. and Forsyth, A.B. (2004): Topaz from Killiecrankie, Flinders Island, and other Bass Strait islands. *The Australian Gemmologist*, 22(1), 2–9.

Bracewell, H. (1998): Gems around Australia; part 14. *The Australian Gemmologist*, 20(3), 108–111.

Bracewell, H. (2006): Torrington and its gemstones. *The Australian Gemmologist*, 22, 479–484.

Keeling, J. L. and Townsend, I. J. (1988): Gem tourmaline on Kangaroo Island. *The Australian Gemmologist*, 16(12), 455–458.

Payette, F. and Klemm, L. (2011): Gem-quality green and blue tourmaline from a Coolgardie pegmatite, Western Australia. The Australian

Schwarz, D. (1991): Australian Emeralds. *The Australian Gemmologist*, 17(12), 488–497.

Webb, G. (2011): Tourmaline from Kangaroo Island, South Australia in the Australian Museum collection. *The Australian Gemmologist*, 24(8), 199.

CHAPTER 7: QUARTZ
Bracewell, H. (1995): Gems around Australia; part 11. *The Australian Gemmologist*, 19(4) 182–184.

Bracewell, H. (2000): A new deposit of smoky quartz crystals from the Torrington area. *The Australian Gemmologist*, 20, 389–390.

Sutton, P. (2002): Quartz from White Rock quarry, Adelaide, South Australia. *Australian Journal of Mineralogy*, 8(1), 17–28.

CHAPTER 8: CHALCEDONY AND OPAL
Cody, A. and Cody, D. (2008): The Opal Story: a guidebook. Melbourne. pp38.

The Australian Gemmologist: Published papers on Opal (1958–2006). CD ROM (2006).

CHAPTER 9: OTHER GEM MATERIALS
Anon. (1985): Jade in South Australia. Mineral Information Series. Department of Mines and Energy, South Australia.

Beasley, A. W. (1973): Petrified wood. *The Australian Lapidary Magazine*. April 1973, 3–5.

Bracewell, H. (1989): Wave Hill prehnite. *The Australian Gemmologist*, 17(4), 127–128.

Goodwin, M. P. (1998): Manganese minerals from the southern New England district. *Australian Journal of Mineralogy*, 4(2), 101–107.

Sutton, P. (2003): Quartz, prehnite and associated minerals from the Wave Hill area, Northern Territory. *Australian Journal of Mineralogy*, 9(2), 73–80.

Additional resources suggested

The following maps are available from The Victorian Government Bookshop, Level 20, Nauru House, 80 Collins Street, Melbourne.

Geological maps of Victoria: 1:250 000 series
Topographic maps of Victoria: 1:25 000 series

Various regional and tourist maps
http://www.bookshop.vic.gov.au/

Maps and Information on obtaining Miners Rights and Tourist Fossicking Authorities is available from the Victorian Government Department of Primary Industries.
http://www.dpi.vic.gov.au/earth-resources/about-earth-resources/publications-and-resources/online

Victorian localities index

Other Australian states localities index

◇◇◇

General index

◇◇◇◇◇◇◇◇◇◇◇◇◇◇◇◇◇◇◇◇◇◇◇◇◇◇

Glossary

◇◇◇◇◇◇◇◇◇◇◇◇◇◇◇

ACCESSORY MINERAL minor mineral in a rock or ore

ALLUVIUM general term for unconsolidated rocks (clay, mud, sand, gravel) deposited by streams

AMPHIBOLE widespread group of complex silicate minerals (includes hornblende)

AURIFEROUS containing gold

AUSTRALITE black glass button found in Australia resulting from melting of rocks hit by a large meteorite

BASALT dark grey, fine-grained volcanic rock containing calcium-rich feldspar and pyroxene

'BIZIL' misspelling of bezel, the portion of a polished stone above the girdle

BRECCIA sedimentary or volcanic rock made up of angular rock fragments cemented into a finer matrix

CABOCHON a gemstone cut having a domed upper surface and a flat base

CALCAREOUS containing calcium carbonate

CHATOYANT showing a wavy or silky sheen, especially by a gemstone cut as a cabochon, due to fibrous inclusions

CLEAVAGE tendency of a crystal to break in a preferred direction along planes of weakness in the structure

COESITE a mineral, a dense form of silicon dioxide which crystallised under very high pressure

COLLET the metal rim in which a gemstone is set

CONCRETION hard rounded mass of mineral formed by precipitation from watery solution in pores of a sedimentary or volcanic rock

CONTACT ROCK an altered rock formed at the contact between an igneous intrusion and surrounding (usually sedimentary) rocks

CRYPTOCRYSTALLINE term referring to aggregates of microscopic crystals or crystalline fibres.

DICHROISM an optical property of a transparent gemstone in which light is absorbed to produce two colours or shades

DIORITE intrusive igneous rock consisting mainly of amphibole and sodium-rich plagioclase feldspar

DISSOLUTION action of dissolving

DODECAHEDRON a twelve-sided form in the cubic system

EUHEDRAL term describing a well-formed crystal with all its faces intact.

FISSION-TRACK DATING method used to determine age of crystals by counting damage trails caused by fission of uranium atoms within the crystal structure

GEODE a spherical body, often with a central cavity, formed by geological processes, usually in volcanic rocks

GRANITIC general term for intrusive igneous rocks resembling granite

HABIT the general shape of a crystal (e.g. acicular, prismatic, tabular)

HEAVY MINERAL a general term for dense minerals which are concentrated by the action of running water

INLIER an area or group of rocks surrounded by outcrops of younger rocks

KIMBERLITE fragmental igneous rock containing mainly altered olivine, with pyroxene, mica, carbonate and chromite

LAMELLAE thin layers in parallel arrangement, often within crystals

LAPIDARY someone who cuts and polishes gemstones

LHERZOLITE intrusive igneous rock consisting mainly of olivine, with some pyroxene

MAAR explosive volcano with a circular shallow crater and a low rim of ash

MAFIC term for minerals containing magnesium and iron (e.g. olivine, pyroxene) in igneous rocks

METASEDIMENTS general term for metamorphosed sedimentary rocks

MOHS SCALE scale of relative hardness of minerals

MONOCLINE type of fold where sedimentary layers on each side of a flexure remain horizontal

MULLOCK waste rock material from mining

OCTAHEDRON symmetrical 8-sided form in the cubic crystal system

'ORIENTAL AMETHYST' old term for purple sapphire

'ORIENTAL EMERALD' old term for green sapphire

'ORIENTAL RUBY' old term for pale pink sapphire

'ORIENTAL TOPAZ' old term for yellow sapphire

PEGMATITE variety of granite in which crystals of quartz, feldspar, mica and other minerals are larger than usual due to slower cooling in the presence of a fluid

PERCUSSION MARK pit formed on a crystal face by the impact of another crystal

PHANTOM term used for a ghost-like outline of an early crystal enclosed in another

PINACOID a face at the termination of a crystal

PIPE near-vertical, cylindrical body composed of igneous rock or of a massive mineral such as quartz

POINTS sapphire or quartz crystal fragments which are pyramidal terminations

PROSPECTIVITY potential for a economic mineral deposits in a region

PYROXENE widespread group of silicate minerals with a chain-like structure, containing mainly calcium, magnesium and iron

RELIEF surface texture on crystal faces

RESORPTION (MAGMATIC) the dissolving of crystals by magma

RHOMBIC refers to crystal faces on the dodecahedron

RHYODACITE old term for a volcanic rock containing between about 65 and 70% silica and with crystals of quartz and feldspar in a fine-grained matrix

RIBBON STONE jasper containing ribbon-like colour stripes

SCEPTRE term used for an elongated crystal with a bulbous overgrowth at one end

SCHILLER play of iridescent (rainbow) colours made by light refracting from internal surfaces of a crystal

SCORIA deposit of lava fragments which have been frothed up by escape of gas before solidification

SILK (see chatoyant)

SPHERULITIC texture shown by a glassy volcanic rock containing spherical masses of fibrous crystals

STAR SAPPHIRE sapphire crystal containing fine parallel fibres aligned along crystal axes and which produce a star like effect in reflected light.

STRIAE old term for striations

STRIATED term referring to closely spaced parallel grooves on a crystal face or rock surface

SUBDUCTION in plate tectonics, the sliding of one plate of the Earth's crust beneath another

TETRAHEXAHEDRON a cubic crystal form having 24 triangular faces that are arranged 4 to each side of a cube

THUNDER EGG see geode

TILLITE soft sedimentary rock, with an irregular mixture of fine and coarse fragments, deposited by glaciers

TRAPEZOHEDRON a cubic crystal form of 24 faces, each face of which is ideally a four-sided figure having no two sides parallel, or a trapezium

TRISOCTAHEDRON solid shape with 24 faces in the cubic system

TUMBLING method of polishing pebbles by turning them in a rotating container with a abrasive powder

TWIN PLANE surface along which two crystals of the same mineral are structurally connected

VEIN narrow slab-like body of quartz or of an igneous rock such as pegmatite

VOLCANICLASTIC term for deposit consisting of fragments of volcanic rock and crystals formed by explosive volcanic eruption

VUGH (VUG) cavity, usually irregular, in an igneous rock

WASH old term for alluvium, especially material being mined

XENOLITH piece of 'foreign' rock caught up in a magma or igneous rock

MINERAL DATA (for Victorian occurrences)

MINERAL AND VARIETY	GROUP	COMPOSITION	COLOUR RANGE	HARDNESS	CRYSTAL SYSTEM
Almandine	Garnet	iron aluminium silicate	dark red, brownish or purplish red	6.5–7.5	cubic
Andradite	Garnet	calcium iron silicate	green, brown	6.5–7.5	cubic
Anorthoclase	Feldspar	sodium potassium aluminium silicate	colourless to white	6–6.5	triclinic

Beryl
  aquamarine
  heliodor
		beryllium aluminium silicate	
			pale blue
			yellowish
				7.5–8	hexagonal

Calcite		calcium carbonate	white brown	3	rhombohedral
Cassiterite		tin oxide	black, brown, orange, red, yellow	6–7	tetragonal

Corundum
  sapphire
  ruby
		aluminium oxide	
			blue, green, yellow, purple, grey, brown
			pink to red
				9	hexagonal

Diamond		carbon	colourless to yellow	10	cubic
Dravite	Tourmaline	sodium magnesium aluminium borate silicate hydroxide	brown	7	rhombohedral
Dumortierite		aluminium iron oxide borate silicate	royal blue	7	rhombohedral
Elbaite	Tourmaline	sodium lithium aluminium borate silicate hydroxide	green, pink	7	rhombohedral
Forsterite *peridot*	Olivine	magnesium iron silicate	 *pale green*	6–5.7	orthorhombic
Grossular	Garnet	calcium aluminium	green, brown	6.5–7.5	cubic

MINERAL AND VARIETY	GROUP	COMPOSITION	COLOUR RANGE	HARDNESS	CRYSTAL SYSTEM
Olenite	Tourmaline	sodium aluminium borate silicate hydroxide		7	rhombohedral
Oligoclase	Feldspar	sodium calcium aluminium silicate	colourless to white	6–6.5	triclinic
Opal		hydrated silicon dioxide		5.5–6.5	
common			*cream, brown, black yellow*		
precious			*blue, green, purple*		
Pyrope	Garnet	magnesium aluminium silicate	pink to orange	6.5–7.5	cubic
Quartz		silicon dioxide		7	rhombohedral
agate			*white, red, yellow, grey, brown*		
amethyst			*purple*		
carnelian			*red, orange*		
chalcedony			*varied*		
citrine			*yellow*		
rock crystal			*colourless, white*		
smoky			*brown*		
Schorl	Tourmaline	sodium iron aluminium borate silicate hydroxide	black, dark brown, dark blue	7	rhombohedral
Spessartine	Garnet	manganese aluminium silicate	orange to red	6.5–7.5	cubic
Spinel	Spinel	magnesium aluminium oxide	black	7.5–8	cubic
Topaz		aluminium floride silicate	colourless, pale blue	8	orthorhombic
Turquoise		copper aluminium phosphate hydroxide hydrate	pale blue	5–6	triclinic
Zircon		zirconium silicate		7.5	tetragonal
hyacinth			*orange, red, brown*		
jargon			*colourless, pale pink*		

Acknowledgements

◇◇◇◇◇◇◇◇◇◇◇◇◇◇◇◇◇◇◇◇◇◇◇◇◇◇◇◇

*For access to institutional
collections and photographs:*
Argyle Pink Diamonds; Rio Tinto Ltd;
the Natural History Museum in London
(Mike Rumsey and Emily Beech);
the Australian Museum in Sydney
(Gayle Sutherland and Stuart Humphreys);
the South Australian Museum in Adelaide
(Allan Pring and Trevor Peters).

*For providing specimens, photographs,
general information or their expertise:*
Scott and Paul Bennett, Ron Perrin,
Les Hooker, Lynne Selwood, Peter
Woodford, Peter Hoppen, Richard Brew,
Dave Roberts, Pat Jordan, Merv and
Lill Legg, Judy Rowe, Alan Wood,
Murray Thompson, Margaret Day,
Grace and the late John Stewart,
Michael Newnham and James Houran.

For imaging, design and production:
Sarah McCaffrey, Clare McClellan,
Ben Healley, John Broomfield and
Jon Augier for photography in
Museum Victoria (see photo credits);
Patty Brown, publisher at Museum
Victoria Publishing; Elizabeth Dias
at studioether.

Photo Credits

◇◇◇◇◇◇◇◇◇◇◇◇◇◇◇◇◇◇◇◇◇◇◇◇◇◇◇◇

Anonymous: 1.3, 2.5, 3.15 (at the Natural History Museum London); 2.30, 2.33, 2.34 (from Rio Tinto Ltd); 7.21, 8.7 (courtesy of Les Hooker), 8.27, 8.30.
Jon Augier (Museum Victoria): 8.24.
Bill Birch (Museum Victoria): 1.1, 2.10, 3.1, 3.23, 3.48, 4.3, 4.4, 5.3, 5.14, 5.16, 6.4, 7.28, 8.16, 8.21.
John Broomfield (Museum Victoria): 3.55 (upper), 5.20, 5.23, 5.30, 6.15, 6.17, 8.17, 8.18, 8.25, 8.26, 9.6, 9.7, 9.8, 9.9, 9.10, 9.12, 9.14.
Frank Coffa (formerly Museum Victoria): 2.6, 2.7, 2.8, 2.9, 2.12, 2.13, 2.14, 2.16, 2.19, 2.20, 2.26, 3.3, 3.5, 3.6, 3.7, 3.8, 3.9, 3.10, 3.11, 3.18, 3.20, 3.22, 3.25, 3.26, 3.27, 3.28, 3.29, 3.31, 3.32, 3.33, 3.36, 3.38, 3.39, 3.40, 3.41, 3.42, 3.44, 3.45, 3.47, 3.51, 4.1, 5.2, 5.6, 5.7, 5.8, 5.9, 5.10, 5.11, 5.12, 5.13, 5.15, 5.17, 6.2, 6.3, 6.6, 6.7, 6.11, 6.13, 7.5, 7.7, 7.10, 7.12, 7.16, 7.17, 7.22, 7.24, 7.25, 7.26, 8.2, 8.3, 8.4, 8.6, 8.14, 8.20, 8.22, 8.23, 8.28, 8.29, 9.2, 9.3, 9.4, 9.5.
Richard Daintree: 3.34.
Frank Holmes: 3.53.
Ben Healley (Museum Victoria): 3.52, 3.58, 4.5, 4.6, 4.7, 4.10, 4.11, 5.18, 5.24, 5.25, 6.14, 7.14, 7.18, 7.19, 7.20, 7.23, 7.30, 7.31, 7.32, 7.33, 8.8, 8.9, 8.10, 8.11, 8.12, 8.13, 8.15, 8.19.
Caroline Harding (formerly Museum Victoria): 2.23
Dermot Henry (Museum Victoria): 3.46, 7.6.
Stuart Humphreys (Australian Museum): 3.55 (lower), 3.56.
Sarah McCaffrey (Museum Victoria): 2.15, 2.24, 2.32, 2.35, 3.12, 3.13, 3.14, 3.19, 3.37, 3.50, 3.54, 3.57, 3.59, 4.8, 4.9, 5.4, 5.19, 5.21, 5,22, 5.26, 5.27, 5.28, 5.29, 5.31, 5.32, 5.33, 6.8, 6.9, 6.12, 6.16, 6.18, 7.3, 7.4, 7.8, 7.9, 7.11, 7.13, 7.27.
Clare McLellan (Museum Victoria): 6.5, 6.14, 7.15, 8.5, 9.11, 9.13.
Trevor Peters (South Australian Museum): 7.29.
Judy Rowe: 3.49
Jeff Scovil: 2.31.

Preface 1997 edition

✧✧✧✧✧✧✧✧✧✧✧✧✧✧✧✧✧✧✧✧✧✧✧✧✧✧✧✧✧✧✧✧

One of the main aims of this book is to dispel the impression that Victoria is a gemstone-free zone. Many minerals having the requisite properties of beauty, rarity and durability to be called gemstones can be found in the State. Admittedly they are not as prolific and valuable as those found in other Australian gem-fields, an many localities are now depleted or out-of-bounds to collectors. However, some research and planning can still lead to worthwhile, even unexpected, discoveries in Victoria. The book is not meant as a guide to fossicking methods or to the art of lapidary, but it does aim to assist with identification of many of the gem minerals through the use of colour photographs, a feature previous publications on Victorian gemstones have lacked.

A further aim of the book is to encourage the long-term preservation of Victorian gem minerals. Whilst historical collections in museums have been invaluable in preparing the book, very few representative specimens collected since the 1960s by gemstone fossickers have been preserved in public institutions.

While as many gemstone localities as possible are described, the book is not intended to be a field guide. Instead, the emphasis has been placed on the geological origin of each of the gem minerals. This approach has determined the arrangement and titles of the chapters.

Gemstones such as diamonds, sapphires, zircons and olivine, which are typically associated with basic volcanic rocks, are described in Chapters 2-4. The main gem minerals in granitic rocks — topaz, tourmaline and beryl — are discussed in Chapter 5. The next three chapters deal with garnets and the many varieties of quartz, which are found in more diverse geological environments. Ornamental minerals and rocks are described in the last two chapters. Within each chapter, the history of discovery, the main localities, the characteristics of the minerals and their geological origins are presented.

It could be said that this book is either 130 years or 30 years too late. If written in the 1860s, it would have benefited from both the availability of gem minerals found at the height of the gold rushes and the enthusiasm of the experts of the day. If published in the 1960s, it would have been able to incorporate more of the local knowledge, so rarely put in print, of fossicking sites visited by gem collectors. It may also have helped to prolong the hobby of gem fossicking and led to new discoveries. Perhaps the great advantage of writing the book in the 1990s is that new ideas on the link between gem minerals and geological processes in eastern Australia are now being discussed. Victoria's long-forgotten gemstones may yet play an important role in this debate.

Bill Birch and Dermot Henry

Acknowledgements 1997 edition

At the outset, we wish to express our gratitude to the Royal Society and the Mineralogical Society for having the trust in us to produce this book. We were certainly not aware of the scope of the project when we started the research several years ago, but have found it to be both challenging and satisfying. The book could never have been completed without assistance from many people who have been involved, in one way or another, with Victorian gemstones. We relied heavily on obtaining information on collecting sites, access to properties, loans from collectors and institutions, and advice from experts, both professionals and hobbyists. Amongst those who kindly loaned or donated specimens are Fred Kapteina, David Vince, Karl Schultz, Tom Marsh, Millie Howell, Dick Nicholls, Joan Derosin, Peter Schlemme, Ross Kitzelmann, Joyce Mibus, Joe Francese, Neil and Yvonne Smith, Fritz and Margaret Bach, Judy Rowe, Edith Oakes, Tony Forsyth, Frank Robinson, Peter Newcombe, David Roberts, John Stewart, Michael Stott, Alan Webb, Ed Richards, Cyril Kovac, Tony Fraser, Ada Donaldson and Barbara Francis. We were particularly pleased to be able to examine and photograph several Beechworth and Eldorado diamonds in private hands, and express out thanks to the owners.

The scarcity of faceted or polished Victorian gemstones in the Museum of Victoria's collections loomed as a problem. This was greatly alleviated by the generosity of many of the above-named collectors. We also express our appreciation to Peter Day and Scott Langford, whose faceting skills were tested to the limit in cutting small sapphires, zircons, garnets and olivines from the Museum's collections. Their efforts have made a great contribution to the book.

Valuable information was provided by Ron Amess, Bruce Campbell, Len Cram, Dorothy Norton, Ken Ellis, Brian Davis, Murray Ellis, Bing and Mary Kittelty, Loretto Redfern, Eric and Helen Humphries, Ray Billingsley, Rodney Start, Trevor Pickering, Les Brent, Geoff Mosley, Mark Keppell and Peter Temby. We are especially grateful to Dick and Wendy Nicholls, Joan Derosin and other members of the Alexandra and Eildon District Lapidary Club for their guidance during several enjoyable days in the field. We are grateful also to those property owners, including Geoff Bichard, Bruce Campbell, Jos Oxley and Mr and Mrs R. Buchanan, who allowed us to visit collecting localities.

We were privileged to gain access to the gem minerals in the collections of the Burke Museum at Beechworth, thanks to assistance from Margaret Carlton and Michael Meek. Dr Andrew Clark and Dr Bob Symes, from the Natural History Museum, London, provided historical information on Victorian gemstones in the Museum's collections and arranged photography of diamonds. Dr Klaus Thalheim (State Museum, Dresden, Germany) provided historical information on early acquisitions of Australian gemstones. Dr Lin Sutherland and Gayle Webb at the Australian Museum were also willing to help with information and advice. Tony Annear (Victorian Gem Clubs Association) and Tom Troiani (Gemmological Association of Australia, Victorian Division) publicised the project through their organisations' newsletters.

At the Museum of Victoria, Dr Tom Darragh provided leads on historical details, in particular many of those published in newspapers, and also read drafts of the chapters on diamonds and sapphires. We must make special mention of the photographic skills of Frank Coffa and John Broomfield, whose colourful output has enlivened the text. Dr Alan Beasley (Curator Emeritus) gave helpful advice on former collectors and localities. We are also grateful for the assistance of staff at the State Library of Victoria and the Victorian Parliamentary Library in tracking down historical illustrations.

The assistance of Michael Bush and Ian McHaffie in our search for mining and mineral references at the Victorian Department of Natural Resources and Environment is much appreciated. Dr Tim McConachy (Rio Tinto Limited) gave permission for descriptions of diamonds in an unpublished report to be reproduced. Ken Nielsen (Kinex Pty Ltd) assisted with information on the Carapooee workings and on field work in the area. The input of Dr Larry Barron (Geological Survey of New South Wales) and Dr Julian Hollis (Golden Hills Mining NL) into their respective chapters is much appreciated. We are especially grateful for the many opportunities we have had over the years to visit sapphire and zircon localities in the Daylesford-Trentham and Ballan regions under Dr Hollis's enthusiastic guidance.

Our special thanks go to Paul Monkivitch and David Fox at Swinburne Design Centre. Paul and David devoted many months to the computer drafting, layout and overall appearance of the book. It belongs to them as much as to us and we hope they enjoyed the experience. We also acknowledge the assistance of Matt Lees at eclipse. Stephen Huxley (Director of the Swinburne Design Centre) and Dr Joyce Richardson (Royal Society of Victoria) gave enthusiastic support for the project throughout.

Finally, we wish to acknowledge the generosity of those companies, organisations and individuals who provided financial support for the book.

Bill Birch and Dermot Henry